中国轻工业"十三五"规划教材

ENGLISH

FOR SHOES

专门用途英语系列教材

鞋靴专业英语

主编 弓太生 万蓬勃

西安交通大学出版社
XI'AN JIAOTONG UNIVERSITY PRESS

内容提要

　　本书是中国轻工业联合会"十三五"规划教材。本书参考国外相关专著、杂志和网页文章等，结合当前制鞋企业的设计、生产、研发和质量管理等活动，以鞋靴起源、足踝结构和鞋楦基础等内容为开篇，进一步拓展到常用制鞋材料、鞋靴造型与结构设计、鞋类生产技术工艺、生产过程品质检验和质量控制以及现代鞋靴研发技术等。本书是国内首部高等教育鞋类专业英语教材，内容丰富、新颖、实用，有助于帮助读者学习鞋类专业英语、提高专业英语的使用能力。

　　本书适合高等院校服装设计与工程专业（鞋靴方向）的教学，也可供鞋服企业及广大鞋服爱好者参考学习。

图书在版编目（CIP）数据

　　鞋靴专业英语 / 弓太生，万蓬勃主编. —西安：
西安交通大学出版社，2024.2
　　ISBN 978-7-5693-3648-1

　　Ⅰ.①鞋… Ⅱ.①弓… ②万… Ⅲ.①制鞋—英语—
高等学校—教材 Ⅳ.①TS943.6

　　中国国家版本馆CIP数据核字（2024）第011212号

书　　名	鞋靴专业英语
主　　编	弓太生　万蓬勃
策划编辑	蔡乐芊
责任编辑	蔡乐芊
责任校对	庞钧颖
封面设计	任加盟
出版发行	西安交通大学出版社
	（西安市兴庆南路1号　邮政编码710048）
网　　址	http://www.xjtupress.com
电　　话	（029）82668357　82667874（市场营销中心）
	（029）82668315（总编办）
传　　真	（029）82668280
印　　刷	西安五星印刷有限公司
开　　本	787 mm×1092 mm　1/16　印张　14.375　字数　306千字
版次印次	2024年2月第1版　　2024年2月第1次印刷
书　　号	ISBN 978-7-5693-3648-1
定　　价	49.00元

如发现印装质量问题，请与本社市场营销中心联系。
订购热线：（029）82665248　（029）82667874
投稿热线：（029）82665371

前　言

我国是世界上最大的鞋靴生产国和出口国。鞋类专业英语对专业人才的培养和涉及外贸业务的制鞋企业都非常重要。为方便院校师生及广大鞋服工作者更好地学习与掌握鞋类专业英语，1999年我们编写了《皮革制品专业英语》一书。根据中国轻工业联合会"十三五"规划教材的编写要求，我们参考了国内外相关专著、杂志和网页文章等内容，结合当前制鞋企业的设计、生产、研发和质量管理等活动过程，在《皮革制品专业英语》的基础上编写了《鞋靴专业英语》一书。本书可用于高等院校服装设计与工程专业（鞋靴方向）的教学，也可供制鞋企业员工以及广大鞋服爱好者参考使用。

全书分5篇18章53课，每课后附有重难点句和中英词汇对照表，最后为总词汇表。第一篇为引言部分，介绍鞋靴发展历程与鞋文化、足踝结构与常见足疾、鞋楦及代表性鞋款等。第二篇为设计部分，由鞋的结构与部件入手，从样板到技术图纸等方面重点介绍了鞋的造型和结构设计，以及鞋靴的规模化定制。第三篇为生产部分。在简述制鞋工艺基本流程的基础上，着重介绍了以提高出裁率为原则的裁断和制帮面部工段、以绷帮成型方法为主的成型工段以及涵盖胶粘、线缝和硫化工艺的底部工段等内容。第四篇为材料和质量控制部分，首先介绍了各种常见的鞋用面料、底料和辅料，然后以专题形式分别介绍了鞋靴产品的质量缺陷、整鞋检测指标及产品质量控制。第五篇为产品研发部分，以目前制鞋技术的发展趋势、计算机辅助设计和仿真技术等内容为切入点，围绕功能鞋靴，介绍了有关运动生物力学和基于可持续发展理念的鞋靴产品设计与生产。

本书由陕西科技大学弓太生、万蓬勃编写，弓太生统稿。参加资料收集、整理、校对等工作的还有郭思逸、苗容容、倪亮辰、石玉娇、康路平、张琳、张诗雨、杜美娴、覃义婷、李文博、周芷薇等研究生同学。

本教材能顺利出版，得益于曾经使用过原自编教材的广大校友和企业技术人员、国内相关专业院校的教师、教材所参考的著作和论文的作者以及西安交通大学出版社的编辑老师的支持和帮助，他们从教材内容的实用性、先进性和专业性等方面都提出了宝贵的建议，谨向他们表示最诚挚的谢意！

　　因编著者水平有限，书中内容难免有不妥之处，敬请读者指正。

主编

目　　录

PART I
Introduction

CHAPTER 1

The History of Shoes

Lesson 1   The Origin of Shoes

There is much evidence that a foot covering was one of the first things made by our primitive ancestors. Records of the Egyptians, the Chinese, and other early civilizations all contain references to shoes. In the evolution of mankind, footwear served various purposes: protection against cold, heat, and injuries. In very ancient times, people used foot coverings the closest available materials: barks, **woven** grass, leaves, or animal skins. These crude coverings were held to the feet with **thongs**.[1] From these primitive beginnings developed the 3 standard kinds of footwear we have today: the **sandal**, the shoe, and the boot.

In hot countries sandals were made from woven **palms**, grass, or plant fibers and attached to the foot with toe loops. In cold countries shoes were made from animal skins to give better protection and keep the feet warm. A piece of treated skin with holes punched around the edge was put under the foot and laced with a leather strap that **acted as** a drawstring to hold the shoe in place around the foot.[2] This was an early example of the **moccasin**.

The history of shoes appears to start during the Middle **Paleolithic** period approximately 40,000 years ago. Because shoes are made of materials that do not survive well in the fossil record, there are no records of early Stone Age shoes. Much of what we know about early footwear is **speculative**, although preserved footprints can indicate the presence of footwear.

Within the Upper Paleolithic layers of the cave site of Grotte de Fontanet in France, a **footprint** shows that the foot had a moccasin-like covering on it.[3] Skeletal remains from the Sunghir Upper Paleolithic sites in Russia (ca. 27,500 BP) appear to have had foot protection. Earlier evidence for shoe use is based on **anatomical** changes that may have been created by wearing shoes.[4] The earliest evidence of this toe **morphology** noted to date is at the Tianyuan 1 cave site in Fangshan County, China, about 40,000 years ago.

Few early shoes have survived. **Fragments** of Bronze Age footwear have been found in **excavations** but not enough to determine styles. But from the Roman times onwards many shoes have survived suggesting that there were many more shoe styles than one would expect.

The oldest known shoes include a pair of 10,000-year-old sandals found in Oregon, made of grass and **sagebrush** bark woven into a sole and straps, and a pair of 9,000-year-old sandals from California, also made of vegetable matter. The oldest shoes made of animal skins were found with the "Ice Man," a 5,300-year-old fossil from the Tyrolean Alps. These shoes were stuffed with **straw** and **moss** for warmth and were similar to simple **one-piece** moccasins found in most northern cultures around the world.

Along with evolution, the techniques of manufacture got more refined: weaving, **stitching**, carving, **tanning**, etc.[5] In addition, the functional demands for footwear got more demanding and the skills to manufacture these goods evolved. Soon, refined footwear became a status for the wealthy and powerful as well as for **rituals**, **rites**, and religion. Further along with the development of the sense of beauty, footwear became the object of skillful **artisans** and started to **underlie** the waves of fashion.

重点及难点句

1. In very ancient times, people used foot coverings the closest available materials: barks, woven grass, leaves, or animal skins. These crude coverings were held to the feet with thongs.

远古时期，人们将树皮、草编物、树叶或动物皮等随手可得的材料作为裹足物，并用皮带将这些粗糙的裹足物固定在脚上。

2. A piece of treated skin with holes punched around the edge was put under the foot and laced with a leather strap that acted as a drawstring to hold the shoe in place

around the foot.

将一块经过处理的兽皮在其边缘上打孔并置于足底，从孔中穿过一根皮条作为拉绳，将（兽皮）鞋子固定在脚上。

3. Within the Upper Paleolithic layers of the cave site of Grotte de Fontanet in France, a footprint shows that the foot had a moccasin-like covering on it.

在旧石器时代晚期洞穴遗址——法国弗塔奈特石窟中，一个脚印显示出这只脚上穿着类似莫卡辛软皮鞋的裹足物。

4. Earlier evidence for shoe use is based on anatomical changes that may have been created by wearing shoes.

人类穿着鞋子的早期证据是基于可能由穿鞋造成的解剖学变化。

5. Along with evolution, the techniques to manufacture got more refined: weaving, stitching, carving, tanning, etc.

随着进化，制造技术变得更加精细，如：编织、缝纫、雕刻、鞣制等。

词 汇

woven：编，织

sandal：凉鞋

acted as：充当……

moccasin：莫卡辛软皮鞋（原为美洲土著所穿），皱头式软帮鞋，包子鞋

Paleolithic：旧石器时代

footprint：脚印

morphology：形态学

excavation：挖掘

straw：稻草

one-piece：一体式的

tan：鞣制

rite：典礼

underlie：构成……的基础

thong：（固定用）皮条，皮带

palm：棕榈树

speculative：推测的

anatomical：解剖的

fragment：碎片

sagebrush：灌木蒿

moss：苔藓

stitch：缝纫

ritual：仪式

artisan：手艺人

Lesson 2 Shoes and Social Status

Footwear refers to the coverings of the feet, usually referring to shoes, also socks, leggings, and other coverings. The oldest forms of footwear were worn for protection, but as cultures became more **sophisticated** and societies became ranked, footwear came to be used to mark the social status as well.[1]

Feet and footwear carry not only the weight of the body, but a great deal of symbolic, social, and cultural weight as well. How we view and treat the foot, the kinds of footwear that we wear, and how we view our footwear tell us a great deal about society and culture. Feet, both bare and **shod**, are linked to our ideas about gender, sexuality, class, and culture. Thus, we can read, through the history of footwear in a given society, the evolution of that society's ideas about men and women, the working classes and the **elites**, and work and leisure. Shoes ultimately signify individual identity, group affiliation, and social position.[2]

In societies in which shoes are worn by many or most people, going barefoot is generally a sign of very low status. Slaves in ancient Rome, for example, went barefoot, as did the poor, while Roman citizens would never be seen outdoors unshod. Bare feet were also associated with poverty during the Middle Ages in Europe, and slaves in the New World often went barefoot as well. In art in cultures around the world, showing a person without shoes often indicates poverty, but it could also indicate another form of lesser status. Roman women, for instance, were often shown barefoot next to Roman citizens in shoes as a way of indicating their lack of freedom and mobility.

In all ranked and stratified societies, one's social position is reflected in the type of clothing and footwear that a person wears. For instance, elites in Africa, Asia, and Europe have traditionally worn much more **elaborately** made and **decorated** shoes than the masses, with royal family members wearing the most elaborate of all. In general, elites throughout history have tended to wear shoes that are less **sensible** and comfortable than those worn by the working classes because working people need to wear shoes that allow them to work, while the wealthy wear often choose footwear to signify status.[3]

In addition, the foot itself often signifies class position.[4] For example, small feet have been desirable attributes for women in societies around the world. Large feet are not only seen as unfeminine but they signify peasants and other people who have to work for a living, and thus shoe styles have long been developed to make women's feet appear smaller and **daintier**. At the same time, those shoes could only be worn by upper-class women because of the discomfort and the lack of utility.[5] Bound feet in history were desirable partly because a woman with bound feet could not work (and often could not walk), demonstrating her high status and wealth.

重点及难点句

1. The oldest forms of footwear were worn for protection, but as cultures became more sophisticated and societies became ranked, footwear came to be used to mark social status as well.

最古老的鞋子是为了起保护作用，但随着文化变得更加复杂，社会变得等级分明，鞋子也开始被用来象征社会地位。

2. Shoes ultimately signify individual identity, group affiliation, and social position.

鞋子最终象征着个人身份、群体归属和社会地位。

3. In general, elites throughout history have tended to wear shoes that are less sensible and comfortable than those worn by the working classes because working people need to wear shoes that allow them to work, while the wealthy wear often choose footwear to signify status.

一般来说，相较于工人阶级，历史上的精英人士倾向穿着的鞋子更不实用，也不舒适，因为劳动人民需要穿着便于工作的鞋子，而富人穿着的鞋子通常是用来表明地位的。

4. In addition, the foot itself often signifies class position.

此外，脚本身往往显示阶级地位。

5. At the same time, those shoes could only be worn by upper-class women because of the discomfort and the lack of utility.

　　同时，只有上流社会的女性才能穿这些鞋子，因为它们穿起来不舒服，而且缺乏实用性。

词　汇

sophisticated：复杂的，精致的 　　　shod：穿鞋（或靴），穿着……鞋的

elite：上层人士，尖子，精英 　　　　elaborately：精巧地

decorated：装饰的，修饰的

sensible：理智的，合理的，朴素而实用的

daint：小巧漂亮的

Lesson 3　Typical Shoes from Around the World

People in different countries often have quite different shoes, whether it be the style, the materials they are made from, or the reasons why people wear them. Highlighted are some of the shoes that can be found in other countries. Think about how you would feel wearing a **bath clog** from Turkey!

1. America's cowboy boot of 20th century

America's cowboy boot was developed in the United States around 1867 when the first big cattle drives began from Texas to Kansas.

The earliest boots, influenced by the **riding gear** of the Mexican **vaquero**, were made by bootmakers in Kansas. Usually black or dark brown they were not fancy but very practical. Their narrow toes slipped easily into **stirrups**, the reinforced steel arches helped brace the feet, **underslung** heels helped the feet stay in the stirrups if the horse stopped suddenly and tall boot tops protected against chafing and brushes with cactus and rattlesnakes.[1]

Spurs were attached to the heels and while they helped to prod the horse to gallop, they served no real purpose while the cowboy walked around town, except to announce his presence with a loud, satisfying jangle.

Figure 1　Cowboy boot

2. Native American footwear

Moccasins are usually associated with Native North Americans. The word moccasin comes from the East Algonquian dialect.

Two main types can be found: the hard sole which has a separate sole and upper, and the soft sole where the sole and upper are cut from one piece of leather.[2] The job of making and repairing them was done entirely by women. Girls started their training at eight years old.

Before contact with Europeans, porcupines, and bird **quills** were used as decorations until European traders arrived with glass beads.

Described by one 19th-century traveler as the pride of the "Indian wardrobe", moccasins were also given away as signs of friendship or traded for food.

Figure 2   Native American footwear

3. Women's bath clogs, Turkey, 1850–1900

Made of wood and inlaid with mother of pearl, this type of platform clog was worn by Turkish women in bathhouses to protect their feet from the wet floors.

Figure 3   Women's bath clogs

4. Men's leather sandals, Ashanti, Ghana, Africa, early 20th century

Sandals have been the dominant footwear in the hot climate regions of Africa, Asia, and the Americas for centuries. After all, sandals are ideal for the heat, since their firm soles protect the feet from scorching surfaces, while the **straps** allow air to circulate freely and sand to fall out easily. The Ashanti is an agricultural people who live in the southern part of Ghana.

Figure 4 Men's leather sandals

5. Men's "sabots", France, 1920–1940

The French call clogs "sabot". This carved wooden pair is painted in yellow, black, and red to resemble a buttoned shoe.

Figure 5 Men's "sabots"

6. Women's "getas", Japan, c. 1963

High-platform wooden "getas" were traditionally worn by the highest class of Japanese geisha.

A very practical shoe they keep the **hem** of kimonos out of dirt. Before entering a house, they are slipped off as they are considered unclean, and outdoor shoes are never to be worn indoors. If it is cold, cotton socks called "**tabi**" are worn with a toe divide for the thong.

Figure 6   Women's "getas"

7. Man's leather boot with a blue tassel, Mesopotamia (now Iraq), late 19th century

This is a traditional shoe that has been worn in Mesopotamia (now Iraq) for over 4, 000 years. Rock carvings by the Hittites have been discovered showing local people wearing a similar style.

Figure 7   Men's leather boot with a blue tassel

8. Men's camel hide slippers, Morocco, 1900–1950

Embroidered with a spiral pattern in red and grey cotton. Notice how the backs have been squashed down. In the Islamic faith, shoes are removed before entering a mosque, so this type of shoe can easily be slipped off and on.

Figure 8   Men's camel hide slippers

9. Toe peg sandals, India, 1875–1899

Since cow leather was forbidden within the Hindu religion, sandals were made from wood, ivory, or metal. They were called **padukas**, **chakris** or **kharrows**.

Figure 9   Toe peg sandals

10. Clog sandals, Nigeria, Africa, 1924

These sandals have a raised wooden sole covered with carved geometric patterns and leather straps.

Figure 10   Clog sandals

重点及难点句

1. Their narrow toes slipped easily into stirrups, the reinforced steel arches helped brace the feet, underslung heels helped the feet stay in the stirrups if the horse stopped suddenly and tall boot tops protected against chafing and brushes with cactus and rattlesnakes.

尖窄的脚头使靴子很容易滑入马镫，钢加固的足弓部位有助于支撑双脚，如果马突然停下来，矮鞋跟有助于靴子"卡"在马镫里，长靴筒可以防止来自仙人掌和响尾蛇的伤害。

2. Two main types can be found: the hard sole which has a separate sole and upper, and the soft sole where the sole and upper are cut from one piece of leather.

主要有两种款式：硬底款是帮底分离的，而软底款的鞋底和鞋帮由一整块皮革组成。

词　汇

bath clog：木质浴室鞋，木底鞋，木屐

riding gear：马具　　　　　　　　vaquero：牛仔，牧人

stirrup：马镫　　　　　　　　　　underslung：重心低的

quill：大翎毛，羽茎，（豪猪、刺猬的）刚毛

strap：带条，绊带　　　　　　　　sabot：木屐，木鞋，木底皮鞋

getas：日式木屐　　　　　　　　　hem：下摆，折边，底边

tabi：（大拇指单独分开的）日式厚底短袜

toe peg sandals：夹趾凉（拖）鞋　　padukas：木质夹趾拖鞋

chakris：象牙质夹趾拖鞋　　　　　kharrows：金属质夹趾拖鞋

CHAPTER 2

The Foot Structure and Anatomy

Lesson 1   Skeletal System

The human feet are very complex. Each foot consists of 26 bones; 33 joints; muscles, tendons, and ligaments; and a network of blood vessels, nerves, skin, and other surrounding soft tissues. These components work together to create a complex flexible structure to provide the body with support, balance, and mobility. The human feet combine mechanical complexity and structural strength. The feet can sustain large pressure and provide flexibility and resiliency.

In order to have a good understanding of the foot, **directional terms** are very useful. The inner side or big toe side of the foot is called the medial side, while the outer side or little toe side is called the lateral side. Distal is away from the trunk, or origin— in this case away from the center of the foot. Proximal is the nearest to the **trunk** or origin—in this case, it means near the center of the foot.[1] **Anterior** is the front side while **posterior** is the back or rear side.

There are 26 bones in each human foot (Figure 1). Considering that the number of bones in the body is 206, both feet make up one quarter of the bones in the entire human body. The bones can be divided into three groups according to their locations and functions. Seven tarsals or tarsus form the ankle as the connection of the foot and leg. Five metatarsals or metatarsus form the medial side to the lateral side. They are numbered I–V. Fourteen phalanges are the bones of the toes. There are three for each toe, except the big toe, which has two.

1. Tarsus

The tarsus consists of **calcaneus** (1), **talus** (2), **navicular** (3), **cuboid** (7), and three cuneiforms (4–6) (Figure 1). The calcaneus, also called the heel bone, is the largest bone of the foot. The talus, which sits on the calcaneus, forms the ankle joint together with the calcaneus by articulating above the fibula and tibia (the two leg bones). It also articulates forward with the navicular on the medial side.[2] The pressure and impact generated by body weight or walking are transmitted to the ground mostly through these two bones.

The navicular is boat-shaped when seen on the medial side of the foot. It is the connection between the talus and cuneiforms. On its medial surface, there is a prominent **tuberosity** for the attachment of the tibialis posterior **tendon**, which may abrade with the inside surface of the shoe and cause foot pain if it is too large.[3]

The cuboid is located between the metatarsal IV and V and the calcaneus, which also articulates with the navicular and the **lateral** cuneiform on its medial side. There are three cuneiforms, named the medial cuneiform (4), the intermediate cuneiform (5), and the lateral cuneiform (6), respectively (Figure 1). They are also called the first cuneiform, the second cuneiform, and the third cuneiform sometimes. The three cuneiforms articulate with the navicular behind and with the bases of the medial three metatarsals in front.

2. Metatarsus

The five metatarsals, which are numbered I–V from the **medial** side to the lateral side, are the connection of the tarsus and phalanges (Figure 1). Each metatarsal has a proximal base connected to the tarsus, a slim shaft, and a **distal** head near the digits. The sides of the bases of metatarsals II–V also articulate with each other. The lateral side of the base of metatarsal V has a prominent tuberosity, which projects posteriorly and is the attachment site for the tendon of the **fibularis brevis muscle**.[4]

3. Phalanges

The 14 phalanges form the toes. The great toe (first toe) has just two phalanges, while the other four have three each: the **proximal** phalanx, the middle phalanx, and the distal phalanx. Each **phalanx** has a base, a shaft, and a head (Figure 1). As discussed, the distal side is away from the center of the foot; hence, the distal phalanx is at the toe tip, while the proximal phalanx is the one near the metatarsals.

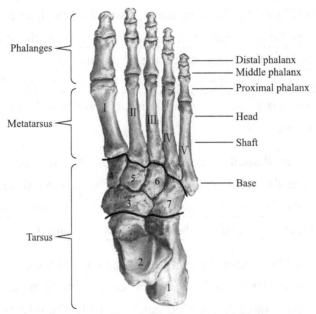

Figure 1　Top view of the skeletal foot

Note: 1. calcaneus; 2. talus; 3. navicular; 4. the medial cuneiform; 5. the intermediate cuneiform; 6. the lateral cuneiform; 7. cuboid

重点及难点句

1. Distal is away from the trunk or origin—in this case away from the center of the foot. Proximal is the nearest to the trunk or origin—in this case it means near the center of foot.

远端是指远离躯干或起点——在本例中是远离足中心。近端指距躯干或起点最近的地方——在本例中指足中心附近。

2. The talus, which sits on the calcaneus, forms the ankle joint together with the calcaneus by articulating above the fibula and tibia (the two leg bones). It also articulates forward with the navicular on the medial side.

距骨位于跟骨上，通过与上面的腓骨和胫骨（两条腿骨）连接与跟骨一起形成踝关节。它也与内侧的舟骨相连。

3. On its medial surface, there is a prominent tuberosity for the attachment of the tibialis posterior tendon, which may abrade with the inside surface of the shoe and cause foot pain if it is too large.

16

　　在其内侧表面有一突出的粗隆，是胫骨后肌腱的附着处，若粗隆过大，可能与鞋的内表面发生摩擦，引起足部疼痛。

　　4. The lateral side of the base of metatarsal V has a prominent tuberosity, which projects posteriorly and is the attachment site for the tendon of the fibularis brevis muscle.

　　第五跖骨底外侧有一个向后方突出的粗隆，是短腓骨肌腱的附着部位。

词　汇

directional term：方向术语 　　　　trunk：躯干

anterior：前部的；前面的　　　　　posterior：后部的；后面的

tarsus：跗骨；踝骨　　　　　　　　calcaneus：跟骨

talus：距骨　　　　　　　　　　　navicular：舟状骨

cuboid：骰骨　　　　　　　　　　cuneiform：楔状骨

tuberosity：骨面粗隆，结节　　　　tendon：肌腱

lateral: 外侧的　　　　　　　　　metatarsus：跖骨

medial：内侧的　　　　　　　　　distal：末梢的，末端的

fibularis brevis muscle：腓骨短肌　phalange：趾骨，指骨

proximal：近端的，近侧的　　　　　phalanx：趾骨，指骨

Lesson 2   Arches of Foot

The foot has two types of arches: **longitudinal arches** and **transverse arches**. The longitudinal arches consist of medial and lateral parts, which distribute the body weight and pressure in different directions with the transverse arches together. The longitudinal arches are essential for supporting and reducing the cost of walking. There are several transverse arches, which are all essential for the foot functions. The foot bones are not arranged in a horizontal plane but in longitudinal and transverse arches supported and controlled by tendons, which absorb and transmit forces and pressure from the body to the ground when standing or moving.[1] When the longitudinal arches are higher than normal, the foot is classified as a **high arch foot**. When the longitudinal arches are low, it is called **flat foot**. Both flat feet and high arch feet do not transmit forces efficiently and hence can lead to foot pains. In addition, these also affect the pressure distribution causing irregular pressure in other parts of the body with long-term problems such as back pains.

1. Longitudinal arch

The foot has two longitudinal arches: medial and lateral. The **medial longitudinal arch** is composed of the calcaneus, the talus, the navicular, the three cuneiforms, and the **metatarsal I–III**, which is higher than the **lateral longitudinal arch** and called the foot arch normally[2] (Figure 1). The bones of the lateral longitudinal arch include the calcaneus, the cuboid, and the **metatarsal IV and V** (Figure 2). Under the medial longitudinal arch, **soft tissues** (such as the **plantar calcaneonavicular ligament**) with elastic properties act as springs.

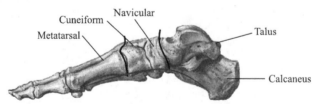

Figure 1   The medial longitudinal arch

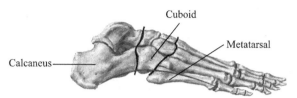

Figure 2   The lateral longitudinal arch

2. Transverse arch

The transverse arch is the arch across the foot from the medial to the lateral side. The shape of the arch can be different at different locations. For illustration, we have shown a **cross section** of the foot at the cuneiform region (Figure 3). The transverse arches together with the longitudinal arches enable the foot in its function of support and locomotion.

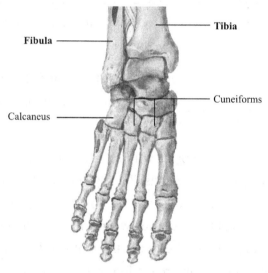

Figure 3   Transverse arch

3. Arch types

The **sole** of the foot generally does not make full contact with the ground due to the two longitudinal arches (medial longitudinal arch, and lateral longitudinal arch) and the transverse arch in each foot.[3] The flexibility of these arches makes walking and running easier, as they provide the necessary shock absorption for the foot. A reliable, valid, and easy way to classify the foot arch type can help discover possible risk factors or potential causes of foot injuries so that appropriate **orthotics** can be prescribed to prevent them.[4] A common categorization is high-arch, normal-arch, or

low-arch (Figure 4).

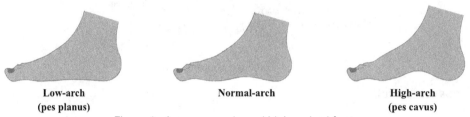

Low-arch   **Normal-arch**   **High-arch**
(pes planus)   **(pes cavus)**

Figure 4   Low-, normal-, and high-arched feet

The low-arched or flat foot has an **imprint** that looks like the sole of the foot. This type of foot is an **overpronated** foot that strikes on the lateral side of the **heel** and has an **excessive roll inward** with a greater risk of soft-tissue damage on the medial side of the foot.[5] Thus, **motion control** and stability are important for a flexible, low-arched foot when selecting a shoe. Often a **straight or mild semi-curve-lasted shoe** will be most appropriate.[6] Low-arched feet are common in infants, children, and about 25% of adults.

The normal-arched foot has an imprint with a **flare**, and the **forefoot** and heel are connected by a wide band. The normal foot would land on the outside of the heel and would roll slightly inward to absorb shock.[7]

The high-arched foot has an imprint without a band or with a very narrow band connecting the forefoot and heel regions. This type of foot is at an increased risk of injuring the bony structures on the lateral side of the foot (**over-supinated**). High-arched feet tend to have smaller areas for **weight-bearing** and also tend to be more rigid, thereby transmitting higher stresses to the foot and leg. Thus, shock absorption is a prime concern for a rigid, high-arched foot when selecting a shoe. Often a **curve-lasted shoe** with a high level of **cushioning** and shock absorption will be most appropriate for those with high-arched feet.[8]

重点及难点句

1. The foot bones are not arranged in a horizontal plane but in longitudinal and transverse arches supported and controlled by tendons, which absorb and transmit forces and pressure from the body to the ground when standing or moving.

足部骨骼不是排列在一个水平面上，而是排列在由肌腱支撑和控制的纵向和横向足弓上，当站立或移动时，肌腱吸收并传递来自身体的力和压力到地面。

2. The medial longitudinal arch is composed of the calcaneus, the talus, the navicular, the three cuneiforms, and the metatarsal I–III, which is higher than the lateral longitudinal arch and called the foot arch normally.

内侧纵弓由跟骨、距骨、舟状骨、三块楔骨以及第 1 至第 3 跖骨组成，且高于外侧纵弓，一般被称为足弓。

3. The sole of the foot generally does not make full contact with the ground due to the two longitudinal arches (medial longitudinal arch, MLA, and lateral longitudinal arch, LMA) and the transverse arch in each foot.

由于每只脚有两个纵弓（内侧纵弓和外侧纵弓）以及横弓，脚底通常不会与地面完全接触。

4. A reliable, valid, and easy way to classify the foot arch type can help discover possible risk factors or potential causes of foot injuries so that appropriate orthotics can be prescribed to prevent them.

一种可靠、有效且简单的足弓类型分类方法可以帮助人们发现可能的风险因素或足部损伤的潜在原因，从而采用适当的矫形器加以预防。

5. This type of foot is an overpronated foot that strikes on the lateral side of the heel and has an excessive roll inward with a greater risk of soft-tissue damage on the medial side of the foot.

这种类型的脚属于过度内翻脚，足跟外侧着地，过度向内翻，足内侧软组织损伤的风险更大。

6. Often a straight or mild semi-curve-lasted shoe will be most appropriate.

通常，直形或半弧形鞋是最合适的。

7. The normal foot would land on the outside of the heel and would roll slightly inward to absorb shock.

正常的脚是脚后跟外侧着地，并微微向内转动以吸收震动。

8. Often a curve-lasted shoe with a high level of cushioning and shock absorption will be most appropriate for those with high-arched feet.

通常具有高度缓冲和减震作用的弧线鞋最适合高足弓的人。

词 汇

arch：足弓

longitudinal arches：足纵弓

transverse arches：足横弓

high arch foot：高弓足

flat foot：扁平足

medial longitudinal arch（MLA）：内侧纵弓

metatarsal I–III：第 1 至第 3 跖骨

lateral longitudinal arch（LMA）：外侧纵弓

metatarsal IV and V：第 4 和第 5 趾骨　soft tissues：软组织

plantar calcaneonavicular ligament：足底跟舟韧带

cross section：横截面

fibula：腓骨

tibia：胫骨

sole：前掌，脚底，鞋外底

orthotics：足部矫形器，矫形鞋垫

low-arch / pes planus：低足弓

normal–arch：正常足弓

high-arch / pes cavus：高足弓

imprint：足印

overpronated：内翻的

heel：脚后跟，鞋后跟

excessive roll inward：过度内旋

motion control：运动控制

straight or mild semi-curve-lasted shoe：直形或半弧形鞋

flare：喇叭形

forefoot：前足

over-supinated：外翻的

weight-bearing：承重

curve-lasted shoe：弧线鞋

cushioning：缓冲

Lesson 3 Foot Disease

The average person takes approximately 10,000 steps per day, which can add up to more than 3 million steps per year. Each step can place 2–3 times the force of your body weight on your feet. You rely on your feet and ankles, so it's important to keep them healthy and pain-free. When an imbalance causes the proper movement pattern of the foot to be disrupted, a number of conditions can result, including the most common conditions outlined below.

1. Achilles tendonitis

Achilles tendonitis is an injury caused by overuse of the large tendon that connects the **calf muscles** to the back of the heel bone. When overused, the achilles tendon can get irritated, painful, stiff, and swollen. Although it is the largest and strongest tendon in the body, it is also the most injury-prone due to its limited blood supply and the enormous stress placed upon it. Achilles tendonitis can worsen if not addressed properly, so address the problem promptly.

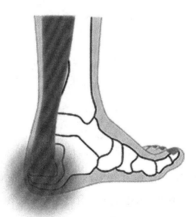

Figure 1 Achilles tendonitis

Cause: There are several factors that can cause achilles tendonitis. The most common cause is over-pronation. **Over-pronation** occurs in the walking process, when the arch collapses upon weight bearing, adding stress on the Achilles tendon.[1] Other factors that lead to Achilles tendonitis are improper shoe selection, inadequate stretching prior to engaging in athletics, a short Achilles tendon, direct trauma (injury) to the tendon, and heel bone deformity.

Treatment & Prevention: Athletes, particularly runners, should incorporate a thorough stretching program to properly warm-up the muscles. They should decrease the distance of their walk or run, apply ice after the activity and avoid any uphill climbs. Athletes should use an orthotic device, **heel cup**, or **heel cradle** for extra support. A heel cup or heel cradle elevates the heel to reduce stress and pressure on the Achilles tendon. The device should be made with light-weight, shock-absorbing materials. An orthotic device can be used to control over-pronation, support the longitudinal arch, and reduce stress on the Achilles tendon.

2. Arch pain—strain

Arch pain (often referred to as arch strain) refers to an inflammation or burning sensation at the arch of the foot.

Figure 2　Arch pain–strain

Cause: There are many different factors that can cause arch pain. A structural imbalance or an injury to the foot can often be the direct cause. However, most frequently the cause is a common condition called **plantar fasciitis**.

Treatment & Prevention: If you suffer from arch pain avoid high-heeled shoes whenever possible. Try to choose footwear with a reasonable heel, soft leather uppers, shock-absorbing soles and removable foot insoles. When the arch pain is pronation related (flat feet), an orthotic designed with a medial heel post and proper arch support is recommended for treating the pain.[2] This type of orthotic will control over-pronation, support the arch and provide the necessary relief.

3. Bunions, bunionettes

Bunions, referred to **hallux valgus** in the medical community, are one of the most common forefoot problems. A bunion is a prominent bump on the inside of the foot around the big toe joint. This bump is actually a bone protruding towards the inside of the foot. With the continued movement of the big toe towards the smaller toes, it is common to find the big toe resting under or over the second toe.

Another type of bunion which some individuals experience is called a Tailor's Bunion, also known as a bunionette. This forms on the outside of the foot towards the joint at the little toe. It is a smaller bump that forms due to the little toe moving inwards, towards the big toe.

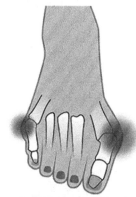

Figure 3   Bunions and bunionettes

Cause: Bunions are a common problem experienced mostly by women. The deformity can develop from an **abnormality** in foot function, or **arthritis**, but is more commonly caused by wearing improper fitting footwear. Tight, narrow dress shoes with a constrictive toe box (toe area) can cause the foot to begin to take the shape of the shoe, leading to the formation of a bunion. Women who have bunions normally wear dress shoes that are too small for their feet. Their toes are squeezed together in their shoes causing the first **metatarsal bone** to protrude on the side of the foot.

Treatment & Prevention: In the early stages of the formation of a bunion, soaking feet in warm water can provide temporary relief. The best way to alleviate the pain associated with bunions is to wear properly fitting shoes. Shoes designed with a high, wide toe box (toe area) are recommended for people suffering from forefoot disorders, such as bunions. Shoes with **rocker soles** will unload pressure to the bunion area.

Orthotics are also recommended for this condition to provide extra comfort, support, and protection. Other conservative treatments include using forefoot products designed to accommodate and relieve bunions such as bunion shields, bunion night splints, and bunion bandages.[3]

4. Calluses

Calluses formation is an accumulation of dead skin cells that harden and thicken over an area of the foot. This callus formation is our bodies defense mechanism to protect the foot against excessive pressure and friction. Calluses are normally found on the **ball-of-the-foot**, the heel, or the inside of the big toe. Some calluses have a deep seated core known as a **nucleation**. This particular type of callus can be especially painful to pressure. This condition is often referred to as **intractable plantar keratosis**.

Figure 4　Calluses

Cause: Calluses develop due to excessive pressure at a specific area of the foot. Some common causes of callus formation are high-heeled dress shoes, shoes that are too small, obesity, abnormalities in the **gait cycle** (walking motion), flat feet, high arched feet, bony prominences, and the loss of the **fat pad** on the bottom of the foot.

Treatment & Prevention: Many people try to alleviate the pain caused by calluses by cutting or trimming them with a razor blade or knife. This is not the way to properly treat calluses. This is very dangerous and can worsen the condition resulting in unnecessary injuries. Diabetics especially should never try this type of treatment. In order to relieve the excessive pressure that leads to callus formation, weight should be redistributed equally with an orthotic. An effective orthotic transfers pressure away from the "hot spots" or high-pressured areas to allow the callus to heal. The orthotic should be made with materials that absorb shock and shear (friction) forces. Women should also steer away from wearing high-heeled shoes.

5. Claw toes

If your toes appear crooked or bent downward you may be suffering from hammer toes, mallet toes or claw toes. A claw toe is a toe that is contracted at the PIP and DIP joints (middle and end joints in the toe), and can lead to severe pressure and pain. Ligaments and tendons that have tightened cause the toe's joints to curl downwards. Claw toes are classified based on the mobility of the toe joints. There are two types— flexible and rigid.

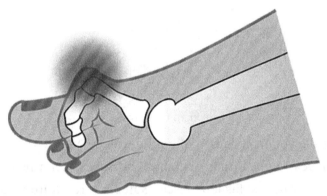

Figure 5   Claw toes

Cause: Claw toes result from a muscle imbalance which causes the ligaments and tendons to become unnaturally tight.[4] This results in the joints curling downwards. Arthritis can also lead to many different forefoot deformities, including claw toes.

Treatment & Prevention: When choosing a shoe, make sure the toe box (toe area) is high and broad, and can accommodate the claw toes. A shoe with a high, broad toe box will provide enough room in the forefoot area so that there is less friction against the toes. Other conservative treatments include using forefoot products designed to relieve claw toes, such as toe crests and hammer toe splints. These devices will help hold down the claw toe and provide relief to the forefoot. Gel toe shields and gel toe caps are also recommended to eliminate friction between the shoe and the toe, while providing comfort and lubrication.[5]

6. Corns

Corns like calluses develop from an accumulation of dead skin cells on the foot, forming thick, hardened areas. They contain a cone-shaped core whose point can press on a nerve below, causing pain. Corns are a very common ailment that usually forms on the tops, sides and tips of the toes. Corns can become inflamed due to constant friction and pressure from footwear.

Figure 6   Corns

Cause: Some of the common causes of corn development are tight fitting footwear, high heeled footwear, tight fitting stockings and socks, deformed toes, or the foot sliding forward in a shoe that fits too loosely. Soft corns result from bony prominences are located between the toes. They become soft due to perspiration in the forefoot area. **Complications** that can arise from corns include bursitis and the development of an **ulcer**.

Treatment & Prevention: Avoid shoes that are too tight or too loose. Use an orthotic or **shoe insert** made with materials that will absorb shock and shear forces. Also avoiding tight socks and stockings to provide a healthier environment for the foot. Diabetics and all other individuals with poor circulation should never use any chemical agents to remove corns.[6]

7. Heel pain

Heel pain is a common condition in which weight bearing on the heel causes extreme discomfort.

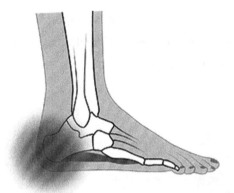

Figure 7   Heel pain

Cause: There are two different categories of heel pain. The first is caused by an over-use repetitive stress. Over-use repetitive stress refers to a soreness resulting from too much impact on a specific area of the foot. This condition, often referred to as "heel pain syndrome", can be caused by shoes with heels that are too low, a thinned out fat pad in the heel area, or from a sudden increase in activity.[7]

Plantar fasciitis, a very common diagnosis of heel pain, is usually caused from a biomechanical problem, such as over-pronation (flat feet).[8] Over-pronation can cause the plantar fascia to be excessively stretched and inflamed, resulting in pain in the heel and arch areas of the foot.

Treatment & Prevention: To properly treat heel pain, you must absorb shock, provide cushioning and elevate the heel to transfer pressure. This can be accomplished with a heel cup, visco heel cradle, or an orthotic designed with materials that will absorb shock and shear forces.

When the condition is pronation related (usually plantar fasciitis), an orthotic with medial posting and good arch support will control the pronation, and prevent the inflammation of the plantar fascia.

Footwear selection is also an important criterion when treating heel pain. Shoes with a firm heel counter, good arch support, and appropriate heel height will be the ideal choice.

8. Over-pronation (flat feet)

Over-pronation, or flat feet, is a common biomechanical problem that occurs in the walking process when a person's arch collapses upon weight bearing. This motion can cause extreme stress or inflammation on the plantar fascia, possibly causing severe discomfort and leading to other foot problems.

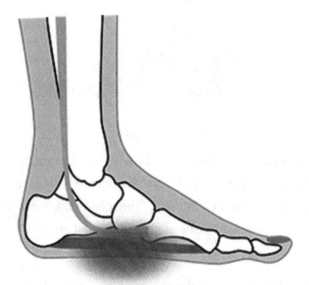

Figure 8   Over-pronation(flat feet)

Cause: There are many causes of flat feet. Obesity, pregnancy or repetitive pounding on a hard surface can weaken the arch leading to over-pronation. People with flat feet often do not experience discomfort immediately, and some never suffer from any discomfort at all. However, when symptoms develop and become painful, walking becomes awkward and causes increased strain on the feet and calves.

Treatment & Prevention: Over-pronation can be treated conservatively (non-surgical treatments) with over-the-counter orthotics. The orthotics should be designed with appropriate arch support and medial rearfoot posting to prevent the over-pronation.

9. Plantar fasciitis

Plantar fasciitis is an inflammation caused by excessive stretching of the plantar fascia. The plantar fascia is a broad band of fibrous tissue which runs along the bottom surface of the foot, attaching at the bottom of the heel bone and extending to the forefoot.[9] When the plantar fascia is excessively stretched, this can cause plantar fasciitis, which can also lead to heel pain, arch pain, and **heel spurs**.

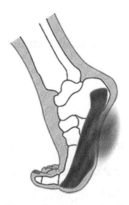

Figure 9   Plantar fasciitis

Cause: The excessive stretching of the plantar fascia that leads to the inflammation and discomfort can be caused by the following: over-pronation (flat feet) which results in the arch collapsing upon weight bearing; a foot with an unusually high arch; a sudden increase in physical activity; excessive weight on the foot, usually attributed to obesity or pregnancy; improperly fitting footwear.

Over-pronation (flat feet) is the leading cause of plantar fasciitis. Over-pronation occurs in the walking process, when a person's arch collapses upon weight bearing, causing the plantar fascia to be stretched away from the heel bone.

Treatment & Prevention: The key for the proper treatment of plantar fasciitis is determining what is causing the excessive stretching of the plantar fascia. When the cause is over-pronation (flat feet), an orthotic with rearfoot posting and longitudinal arch support is an effective device to reduce the over-pronation and allow the condition to heel.[10]

If you have usually high arches, which can also lead to plantar fasciitis, cushion the heel, absorb shock and wear proper footwear that will accommodate and comfort the foot. Other common treatments include stretching exercises, plantar fasciitis night splints, wearing shoes that have a cushioned heel to absorb shock, and elevating the heel with the use of a heel cradle or heel cup. Heel cradles and heel cups provide extra comfort, cushion the heel, and reduce the amount of shock and shear forces placed during everyday activities.

Every time your foot strikes the ground, the plantar fascia is stretched. You can reduce the strain and stress on the plantar fascia by following these simple instructions:

Avoid running on hard or uneven ground, lose any excess weight, and wear shoes and orthotics that support your arch to prevent over-stretching of the plantar fascia.

重点及难点句

1. Over-pronation occurs in the walking process, when the arch collapses upon weight bearing, adding stress on the Achilles tendon.

过度内旋会发生在行走过程中，此时足弓因负重而塌陷，增加了跟腱的压力。

2. When the arch pain is pronation related (flat feet), an orthotic designed with a medial heel post and proper arch support is recommended for treating the pain.

当足弓疼痛与内旋有关（扁平足）时，建议使用设计有内侧足跟支撑和适当足弓支撑的矫正器来治疗疼痛。

3. Other conservative treatments include using forefoot products designed to accommodate and relieve bunions such as bunion shields, bunion night splints, and bunion bandages.

其他保守治疗包括使用前足产品来适应和缓解拇囊炎，如拇囊炎护盾、拇囊炎夜夹板和拇囊炎绷带。

4. Claw toes result from a muscle imbalance which causes the ligaments and tendons to become unnaturally tight.

爪状趾是由于肌肉失衡导致韧带和肌腱变得不自然地紧绷所引起的。

5. Gel toe shields and gel toe caps are also recommended to eliminate friction between the shoe and the toe, while providing comfort and lubrication.

也推荐使用凝胶护趾板和凝胶护趾帽，以消除鞋和脚趾之间的摩擦，它们同时也可带来舒适和润滑。

6. Diabetics and all other individuals with poor circulation should never use any chemical agents to remove corns.

糖尿病患者和所有其他血液循环不良的人都不应该使用任何化学药剂来去除鸡眼。

7. This condition, often referred to as "heel pain syndrome", can be caused

by shoes with heels that are too low, a thinned out fat pad in the heel area, or from a sudden increase in activity.

这种情况通常被称为"脚跟疼痛综合征"，可能是由于鞋子的鞋跟过低、足跟部位脂肪垫变薄或活动突然增加所造成的。

8. Plantar fasciitis, a very common diagnosis of heel pain, is usually caused from a biomechanical problem, such as over-pronation (flat feet).

足底筋膜炎是一种非常常见的足跟疼痛，通常由生物力学问题引起，如过度内旋（扁平足）。

9. The plantar fascia is a broad band of fibrous tissue which runs along the bottom surface of the foot, attaching at the bottom of the heel bone and extending to the forefoot.

足底筋膜是一条沿着足部底表面延伸的宽的纤维组织带，附着在跟骨底部并延伸到前足。

10. When the cause is over-pronation (flat feet), an orthotic with rearfoot posting and longitudinal arch support is an effective device to reduce the over-pronation and allow the condition to heel.

当原因是过度内旋（扁平足）时，（带）后足支撑和纵向足弓支撑的矫形器是一种有效的装置，以减少过度内旋并使足底筋膜回归到正确的后跟位置。

词 汇

achilles tendonitis：跟腱炎   calf muscle：小腿肌肉

over-pronation：过度内旋，外翻

heel cup：后跟垫，跟杯，跟部凹形插件

heel cradle：后跟支架   plantar fasciitis：足底筋膜炎

bunion：拇囊炎   bunionette：小趾囊炎

hallux valgus：拇趾外翻，拇外翻   abnormality：异常，畸形，变态

arthritis：关节炎   metatarsal bone：跖骨

rocker sole：摇杆底   callus：皮肤的硬结，老茧，胼胝

ball-of-the-foot：脚掌   nucleation：成核

intractable plantar keratosis：顽固性足底角化症

gait cycle：步态周期   fat pad：脂肪垫

hammer toe：锤状趾   mallet toe：槌状趾

claw toe：爪状趾   corn：鸡眼

complication：并发症   ulcer：溃疡

shoe insert：鞋垫   plantar fasciitis：足底筋膜炎

heel spur：足跟刺痛

CHAPTER 3

The Shoe-Last

Lesson 1   Shoe-Last Materials

The shoe is expected to wear well, feel well, keep its shape with wear, retain its style character, tread properly, allow for reasonable foot freedom, maintain both foot and shoe balance, and remain structurally intact.[1] These features are not always dependent on the quality of materials or components, or the manufacturing process. The design and multiple dimensions of the last provide the basis for the above.

Studies show over 90% of people have different sized feet.[2] Despite feet being the same length other dimensions vary with the individual foot. Activity of the foot will also change the shape of identical feet in accordance with the structure and function of the **appendage**. Because distribution or proportions of foot mass differ with individuals, linear measurement such as length and breadth of the foot alone, are inadequate.[3] The modern last is made in three dimensions although it is not a direct replica of the foot. The last maker may take up to 35 measurements before the model last can be made.

The shoe-last development depends entirely on the skill and artistry of the model maker who shapes and sands a shoe-last until it looks right. The **model shoe-last** is then graded in order to make shoe-lasts of different sizes. The sized shoe-lasts are used to make shoes to fit the entire selected population. In fact, the shoe design should start with an understanding of the foot and then the shoe-last should be designed.

Shoe-lasts, used extensively in the making of footwear, are forms that are made to varying degrees in the shape of the human foot, depending on their specific purpose.[4] They come in many styles and sizes, depending on the exact job they are designed for. They range from simple one-size shoe-lasts used for repairing soles and heels, to hard-wearing shoe-lasts used in modern mass production, and to custom-made shoe-lasts used in the making of customized footwear. The shoe-last must represent the anatomical information of the foot at the same time giving the finished shoe a pleasing and fashionable appearance.

Shoe-lasts are made from a number of materials. These materials should have shape retaining characteristics and the shoe-last shape must not change with heat, humidity, and other environmental factors. The material used to make modern shoe-lasts must be strong enough to withstand the forces of mass production machinery, such as that applied by the **pullover machines** when bottoming the shoe, and must also be able to hold **tacks** (a type of nail), which are used to hold shoe parts together temporarily before the sole is added.[5] The most commonly used materials for making shoe-lasts are wood, high-density polyethylene (HDPE), and aluminum.

Wooden shoe-lasts are mostly used for **customized** shoemaking. These shoe-lasts are expensive to make and also have low durability. These shoe-lasts may swell or shrink with temperature and humidity. But HDPE is used extensively for all kinds of shoe-lasts. The shoe-lasts made of HDPE are less expensive and can be recycled for new shoe-lasts. The shoe-lasts made of aluminum are used for **solid** or **scoop shoe-lasts** but are very expensive. These shoe-last shapes are also dependent on temperature. Aluminum shoe-lasts are used for making rubber boots.

重点及难点句

1. The shoe is expected to wear well, feel well, keep its shape with wear, retain its style character, tread properly, allow for reasonable foot freedom, maintain both foot and shoe balance, and remain structurally intact.

鞋子应耐磨、脚感好，穿着时能保持形状和款式特征，正常行走并使脚有合理的自由度，能保持脚和鞋的平衡及鞋子的结构完整。

2. Studies show over 90% of people have different sized feet.

研究表明，90%以上的人双脚尺寸不一致。

3. Because distribution or proportions of foot mass differ with individuals, linear measurement such as length and breadth of the foot alone, are inadequate.

由于足部质量的分布或比例因人而异，仅对足部长度和宽度等线性测量是不够的。

4. Shoe-lasts, used extensively in the making of footwear, are forms that are made to varying degrees in the shape of the human foot, depending on their specific purpose.

在制鞋过程中广泛使用的鞋楦在不同程度上与人脚形状相似，这取决于其特定用途。

5. The material used to make modern shoe-lasts must be strong enough to withstand the forces of mass production machinery, such as that applied by the pullover machines when bottoming the shoe, and must also be able to hold tacks (a type of nail), which are used to hold shoe parts together temporarily before the sole is added.

现代鞋楦的制作材料必须足够坚固，能够承受规模生产时的机械力，如底部工段中套楦机所施加的力，并且还能够固定绷帮钉，在合底之前绷帮钉将鞋部件临时固定在楦上。

词 汇

appendage：下肢

pullover machine：套楦机

customized：定制的

scoop shoe-last：活盖鞋楦

model shoe-last：母楦

tack：绷帮钉

solid：实心的

Lesson 2   Types of Shoe-Lasts

It is apparent that the proper fit can be achieved if the shoe is shaped like the foot; however, due to the traditional method of shoemaking, the shoe shape and style are dependent on the shoe-last.

The shoe-last is a **reproduction** of the approximate shape of the human foot. A shoe when properly constructed using a well-designed shoe-last furnishes support and protection without undue pressure, binding, or **constriction** at any point.

The shoe-last is the heart and the single most important element of the shoe. It is the most scientific and complex part of the whole shoemaking process and it is the foundation upon which much of the shoe related foot health depends.

There are many kinds of shoe-lasts used in different shoe industries. The most used types of shoe-lasts are described below.

Solid last: This kind of shoe-last is the simplest and is used for low-heel shoes and sandals.

Figure 1   Solid last

Scoop black last: These lasts are used for the **manual** shoe production. The shoe-lasts have a **wedge** on the top and can be detached from the main body.[1] The lasts can be easily taken out of the lasted shape by removing the wedge.

Figure 2   Scoop black last

Hinge last: These lasts are used for all kinds of shoe production. The lasts have a fore part and a back part and are connected by a spring. When slipping, the last is **bent** to shorten at the V-cut hinge.[2] Then the last is removed from the shoe without damaging or deforming the back part of the shoe.

Figure 3   Hinge last

Telescopic last: These lasts are similar to the hinge lasts but without a V-cut. The last will slid and reduce in length when slipping.

Figure 4   Telescopic last

Three-piece last: These lasts are mainly used for the forced-lasting boot or reversed slippers in the past.[3] These kinds of shoe-lasts consist of three pieces. The center and the back pieces will be removed when slipping. After the removal of the center and back parts, it becomes easier to remove the front part.

Figure 5   Three-piece last

重点及难点句

1. The shoe-lasts have a wedge on the top and can be detached from the main body.

鞋楦的顶部有楔状楦盖，可以从主体上拆下。

2. When slipping, the last is bent to shorten at the V-cut hinge.

入楦时，将鞋楦后身上扳，（V形切口合并）鞋楦尺寸缩短。

3. These lasts are mainly used for the forced-lasting boot or reversed slippers in the past.

过去，这些鞋楦主要用于靴子的套楦成型或拖鞋的翻（反）绱。

词 汇

reproduction：复制品，仿制品

solid last：整体楦

wedge：楔状楦盖

bent：弯曲

three-piece last：三节楦

constriction：束缚

manual：手工的

hinge last：弹簧铰链楦

telescopic last：伸缩式鞋楦

slipper：拖鞋

Lesson 3 Comparison of Foot and Shoe-Last

Many centuries ago, shoes were made in either of two ways: custom-made by a shoemaker; or the individual made his own for himself or his family. The home-made method was relatively simple. The foot was placed on a slab of leather and a sole was cut from it. A piece of leather or cloth was laid over the top of the foot, cut to fit, then nailed or tacked to the sole. And by repeated experience some women learned to make some quite elegant cloth shoes by this simple method.

The shoemaker followed **pretty much the same** basic method except for much more skill and sophistication. He started with **foot tracing** (sometimes even a foot imprint in clay). But measurement of the foot "mass" was also important. And this he did with the "**hand span**" method, determining the **girth** at the **ball**, **instep** and elsewhere with various spans of his hand which he "translated" into a last.[1] He probably also used a crude kind of **size stick** for other measurements.

The last is a roughly foot-shaped form made of molded plastic, carved wood, **cast aluminum**, or 3D-printed plastic. Why is the last called the last? The word "last" comes from the old English word "laest", which means "footprint". The first shoe lasts were used by the Greeks and Romans all the way back to 400 BC.

The modern shoe last is not a replica of the human foot. The shoe last is a **generalization** of the human foot with care taken to account for natural **articulation** and volumetric changes as the foot moves and flexes.[2] A high-quality running or walking shoe will have a thoughtfully developed last. A large shoe brand will have the help of shoemakers, **podiatrists**, and **kinesiologists** to develop **biomechanical** ideal shapes.

It is very essential to understand the difference between a shoe-last and a human foot. The shoe-last is for shoemaking while the foot is used for weight bearing and locomotion. The outline of a shoe-last is regular and continuous with a sharp **feather edge** around the **seat** and **forepart** to assist shoe-lasting and gives a clear defined edge to the finished shoe.[3] The foot has no feather edge. The shoe-last increases gradually in height from the feather line, but not in the foot. The shoe-last surface is **smooth** to enhance the appearance of the shoe and to enable the upper to be mounded more easily to shape. The surfaces of the feet are irregular and vary with individuals.

The shoe-last is hard and firm, while the foot is softer and more flexible. The foot has separate toes, while the toe end of a shoe-last is solid. The **back** curve is greater on the shoe-last to help the shoe grip the foot. The heel height is present in the shoe-last but not on the foot. The front part of the shoe-last is thinner to help the shoe to grip the foot around the quarters (front part of shoe).[4] The shoe-last length is greater than the foot to prevent pressure on the foot. Toe spring is not present on foot but is included on shoe-lasts. Girth and size intervals are regular on shoe-lasts but irregular on feet.[5] The dimensions are identical on a pair of shoe-lasts but rarely identical on a pair of feet.

重点及难点句

1. And this he did with the "hand span" method, determining the girth at the ball, instep and elsewhere with various spans of his hand which he "translated" into a last.

他用"手量"的方法来确定跖趾部位、脚背和其他地方的围度，并将其"转换"到鞋楦上。

2. The shoe last is a generalization of the human foot with care taken to account for natural articulation and volumetric changes as the foot moves and flexes.

鞋楦是对人脚特点的概括，要同时考虑到足部运动和弯曲时的关节和体积变化。

3. The outline of a shoe-last is regular and continuous with a sharp feather edge around the seat and forepart to assist shoe-lasting and gives a clear defined edge to the finished shoe.

鞋楦的轮廓是规则的、连续的，前掌及后跟部位的底边缘清晰，便于绷帮并赋予成鞋清晰的子口线。

4. The front part of the shoe-last is thinner to help the shoe to grip the foot around the quarters (front part of shoe).

楦的前端较廋以使鞋的前帮部位抱脚。

5. Girth and size intervals are regular on shoe-lasts but irregular on feet.

楦的围度和尺寸间隔是有规律的，但脚的则不是。

词　汇

pretty much the same：相差无几

hand span：用手丈量，掌距

ball：拇趾球，跖趾部位

size stick：尺寸测量器，量脚尺

generalization：概括

podiatrist：足科医生

biomechanical：生物力学的

seat：后跟部位

smooth：光滑的，烫平

foot tracing：足印

girth：围度

instep：脚背，跗面

cast aluminum：铸铝

articulation：关节

kinesiologist：人体运动学家

feather edge：楦底棱

forepart：前段，前帮

back：脚背，跗背，背衬

CHAPTER 4

Common Styles of Shoes

Lesson 1   Types of Shoes

There are a wide variety of different types of shoes. Most types of shoes are designed for specific activities. For example, boots are typically designed for work or heavy outdoor use. Athletic shoes are designed for particular sports such as running, walking, or other sports. Some shoes are designed to be worn on more formal occasions, and others are designed for casual wear. There are also a wide variety of shoes designed for different types of dancing. Orthopedic shoes are special types of footwear designed for individuals with particular foot problems or special needs.

1. Men's shoes

Oxfords (Balmorals)—The main feature of these shoes is the I or V-shaped slit in the vamp where laces are attached.

Blüchers (Derbys)—Similar looking as Oxford's, but laces are attached to two pieces of leather that are glued or stitched to vamp.

Monk-straps—Shoes that don't have laces, but instead use buckle and strap to secure the shoe around the foot.

Brogues—They feature decorative **perforations** and separations along the visible edges of material.[1] Previously used primarily in outdoor footwear, brogue perforations can today be found in many types of casual shoes.

Slip-ons—Shoes that don't have any securing mechanism such as lacings or **fastenings**. Most popular shoes of this type are loafers and **elastic-side shoes**.

Plain-toes—Design that doesn't features additional layer of material on the vamp in the area of toes.

Cap-toes—Much more popular design, it features an extra layer of leather that caps the toes.

2. Women's shoes

High-heeled footwear—Most commonly used in formal occasions and social outings. They feature a heel that is typically 2 inches high.

Stilettos—Invented as a high fashion accessory, this type of heel is very popular among women. It features long and very narrow heel posts.

Slingbacks—High heeled shoes that are secured not by over-the-top straps, but ones that go behind the heel.

Mules—Shoes or slippers that have no fastenings around the ankle. They may have small or high heels, and only fastenings are located around the toes and lower part of the feet.

Ballet flats—They are popular in warm environments and feature very low and flat heels.

Court shoes (pumps)—Popular high-heeled shoes that can be easily slipped-on.

3. Unisex shoes

Sandals—Popular throughout history as one of the simplest footwear to be found. They feature a minimal number of straps that leave much of the foot exposed to the air.

Slippers—Used mostly indoors, similar to sandals, only with the vamp that covers one-half of the foot.

Moccasin—Popularized by native Americans, this soft leather shoe has no heel and is usually intended for outdoor use.

Platform shoe—It features very thick soles and heels. It was used in ancient Greece as a sign of wealth, and in recent times as a fashion item.

Boots—Universal footwear for harsh environments, made from leather or rubber.

Slip-on shoe—Casual footwear that features no laces.

Sneakers—Popular casual shoes that are made from leather, canvas and rubber. They were introduced in 1917 and ever since then, they remain extremely popular across the entire world.

4. Athletic shoes

Various sports demand from the user to protect their feet in specific ways. Some of the most popular shoes for such activities are running shoes (they have an emphasis on cushioning), track shoes (featuring metal spikes for additional friction with the ground), climbing shoes (lightweight with rubber **exterior**), skating shoes (modeled to accept skating attachments), cycling shoes (modeled to better interact with bicycle pedals) and **wrestling shoes** (modeled to provide protection and additional traction).

5. Dance shoes

Dance shoes are a specific type of shoes that is used by many dancers. Some of the most popular ones are ballet shoes (made from soft materials), **pointe shoes** (made for ballet dancing, with hardened soles so dancers can stand on the tip of their toes), **ballroom shoes** (created either for traditional dances or for Latin American dances), dance sneakers, tap shoes, tango and flamenco shoes and jazz shoes.

重点及难点句

1. They feature decorative perforations and separations along the visible edges of material.

它们的特征是沿着材料的可见边缘有装饰性的花孔和分隔线条。

词 汇

Oxfords: 牛津鞋　　　　　　Blüchers (Derbys): 外耳式鞋，德比鞋

monk-straps：搭扣鞋，孟克鞋   brogues：布洛克鞋

perforations：孔眼，花孔

slip-ons（slip-on shoe）：便鞋，一脚蹬鞋，无扣便鞋

fastening：紧固件、束紧装置   elastic-side shoes：松紧带鞋

plain-toes：素头式   cap-toes：包头鞋

stilettos：细高跟鞋   slingbacks：后跨带高跟鞋

mules：穆勒鞋，皮拖   ballet flats：软帮平底鞋

court shoes：船鞋   exterior：外观，外部，外貌

wrestling shoes：摔跤鞋   pointe shoes：芭蕾舞鞋

ballroom shoes：舞鞋

Lesson 2 High-Heeled Shoes

Do you really know high heels? Most likely not.

High-heeled shoes, or abbreviated as high heels or merely heels are a kind of footwear that raises the heel of the wearer's foot considerably higher than the toes.

In the fifteenth century, higher heels had been invented by the court tailor in France. From then on, people's craze about them started and never decreased.[1] Not until the late sixteenth century did higher heels become the fashionable products for the nobilities. It was said that in purpose to seem taller, more powerful and more authoritative, Louis XIV ordered his shoemaker to make a four-inch-high heel and painted the heels crimson to show his dignity.

In the 17th century, high heels became essential items for ornamentation. The higher heels at that time were 3 inches high. If you had the chance to stroll on the European streets of the 17th century, you would surprisingly discover that all the people were sporting the exact same high heels, because at that time, that was the only style they could make. From the late 17th century, people tried to make thinner heels, and in order to make it stable, they broadened the top of the heels.[2] In the 18th century, the peak of the heels was down gradually, rather, the high heels ornamented with **ribbons** and **rosettes** had been extremely well-liked. In the 19th century, the Mary Jane fashion was set up, which stored been well-liked for fifty many years. At that time the shoe making methods were higher enough to make different styles.

In the 20th century, higher heels developed quickly. In the nineteen twenties, the women's attitude towards **attire** became newer and open minded, and the ethical rules at that time had been lessened. The designers tried to mix the sandals with the high heels and they turned out to be the sophisticated higher heels sandals for banquets.[3] With the recognition of **peep toe high heels**, **sling backs** also turn out to be popular. Throughout that time, numerous style magazines satirize the sling backs and though it was impolite wore them in the community. Nevertheless, based on the ladies' want for liberation, this point was soon deserted. Until the 1950s, because of the metal nail technique, the designers had been able to style the **stiletto heel**, which was cherished and hated by women until now.[4] At that time when Marilyn Monroe became famous right away for the golden stiletto high heels developed by Salvatore Farragamo, she said, though I don't know who invented high heels, each lady should be grateful to him. The high heels assisted me greatly in my career.

In western countries, high-heeled shoes exist in two highly gendered and parallel tracks: highly fashionable and variable women's shoes with thin long heels, and practical, relatively uniform men's shoes in a riding boot style, with thick, relatively short heels. Heels are often described as a sex symbol for women, and magazines, as well as other media sources that primarily portray women in a sexual way, often do so using high heels. Paul Morris, a psychology researcher at the University of Portsmouth, argues that high heels accentuate "sex-specific aspects of female gait," artificially increasing a woman's femininity. Despite the sexual connotations, heels are considered both fashionable and professional dress for women in most cases, the latter especially if accompanied by a pants suit. Some researchers argue that high heels have become part of the female workplace uniform and operate in a much larger and more complex set of display rules. High heels are considered to pose a dilemma to women as they bring them sexual benefits but are detrimental to their health. The 21st century has introduced a broad spectrum and variety of styles, ranging from the height and width of the heel, to the design and color of the shoe.

重点及难点句

1.From then on, people's craze about them started and never decreased.
从那之后，人们开始狂热追求它，而且这种热情从未减退。

2. From the late 17th century, people tried to make thinner heels, and in order to make it stable, they broadened the top of the heels.
从 17 世纪后期开始，人们尝试制造更细的鞋跟，并且为了使其更加稳定加宽了鞋跟的掌面。

3. The designers tried to mix the sandals with the high heels and they turned out to be the sophisticated higher heels sandals for banquets.
设计师们尝试将凉鞋与高跟鞋混搭，制造出了宴会用的精致高跟凉鞋。

4. Until the 1950s, because of the metal nail technique, the designers had been

able to style the stiletto heel, which was cherished and hated by women until now.

　　直到 20 世纪 50 年代，由于采用了金属钉技术，设计师们才得以设计出细高跟鞋的款式，直到现在女性仍对这种款式又爱又恨。

词　汇

　　ribbon：丝带

　　rosette：花结

　　attire：服装，盛装

　　peep toe high heels：鱼嘴高跟鞋，前空式高跟鞋

　　sling backs：后空式高跟鞋

　　stiletto heel：锥形细高跟，匕首跟

PART II
Design of Shoes

CHAPTER 1

Structure and Part Name of Shoes

Lesson 1   Shoe Structure

Modern shoes have a relatively straightforward structure. Shoes are fashioned out of a sole, insole, upper, **tongue**, heel, **quarter**, and **vamp**.[1] Most shoes also have a lining inside the shoe.

The sole of a shoe is the part of the shoe that lies between the bottom of the foot and the ground. It generally consists of the outsole, which touches the ground, the insole, which lies inside the shoe and supports the foot, and the midsole, which is the materials between the insole and the outsole.

Modern shoes often have **orthopedic** or removable insoles for extra comfort. The midsole is the area of the shoe that modern technology focuses on. Made today of foam, many athletic shoe companies also have their own patented cushioning systems like air cushions or gel that are inserted into the midsole.

The **shank** is a piece of metal found in some shoes, which lies between the sole and the insole and provides support to the arch of the foot. Shanks are necessary in high-heeled shoes.

The heel is attached to the bottom of the outsole, under the heel of the foot. Traditionally made of stacked leather, today heels can be made of wood, cork (for platform shoes), plastic, and steel (for stilettos).

The quarter is the part of the upper that lies behind the heel of the shoe and encapsulates the rear of the foot.[2] The heel counter is a piece of stiff material, above the heel of the shoe at the back of the upper. It lies between the inner lining and the quarter and stabilizes the foot and maintains the shape of the shoe.[3]

The upper is the part of the shoe that attaches to the sole and covers the entire foot. The most traditional material for shoe uppers is leather, followed by cloth. Today, uppers can be made from canvas (as in athletic shoes), leather, cotton, rubber, or a variety of synthetic materials.

The vamp is the part of the upper that covers the front of the foot from the middle to the toe, and where the shoe laces, **buckles, buttons**, or velcro are found.[4] In front of the vamp lies the toe box, which surrounds and protects the toe.

The toe box is the front of the vamp that covers the toes. In some shoes, the toe box is reinforced, as in steel-toe shoes for industrial work, or pointe shoes for ballet dancers. Because narrow feet are often preferred for women, toe boxes are often more narrow than the toes are, causing corns, bunions, and hammer toes to develop.[5]

The tongue is featured in modern shoes that lace or connect on the front of the upper. The tongue is the separate piece of leather or cloth that lies between the two separate pieces of the upper, and over which the laces tie.

Shoes are fastened through a variety of means. Shoes that are fastened with shoe laces, which can be made of materials like cloth, **sinew**, twine or leather, have holes in the shoes for the laces to pass through. Today these are called eyelets. Modern cloth shoelaces have a plastic coating at the tip of the laces called **aglets**, which makes it easier for the laces to pass through the eyelets. Other methods of shoe closure are buttons, **snaps**, buckles, and, today, velcro.

重点及难点句

1. Shoes are fashioned out of a sole, insole, upper, tongue, heel, quarter, and vamp.

鞋子由鞋底、内底、鞋面、鞋舌、鞋跟、后帮和前帮构成。

2. The quarter is the part of the upper that lies behind the heel of the shoe and encapsulates the rear of the foot.

后帮是鞋面的一部分，位于鞋的后部，包裹着脚的后部。

3. It lies between the inner lining and the quarter and stabilizes the foot and

maintains the shape of the shoe.

它（主跟）位于鞋跟上方的鞋后帮处，具有保持鞋形、稳定足部的作用。

4. The vamp is the part of the upper that covers the front of the foot from the middle to the toe, and where the shoe laces, buckles, buttons, or velcro are found.

前帮是鞋面的一部分，它覆盖着直到脚趾的足前部，鞋带、鞋钎、鞋扣或魔术贴都位于鞋面的前帮。

5. Because narrow feet are often preferred for women, toe boxes are often more narrow than the toes are, causing corns, bunions, and hammer toes to develop.

由于女性更喜欢纤细的脚形，所以其鞋头通常比脚趾更窄，因而会导致形成鸡眼、拇囊炎和锤状趾。

词 汇

tongue：鞋舌	quarter：后帮
vamp：前帮	orthopedic：矫形的，骨科的
shank：胫，小腿	buckle：鞋钎，鞋扣
button：鞋扣	sinew：松紧带
aglet：鞋带封头	snap：四合扣

Lesson 2 Shoe Pattern Parts

Shoes consist of considerably fewer main parts than the foot itself, but each is still designed to work with the movement of the foot.

Before starting to design a shoe, it is important to understand each of component. It is also helpful to learn to identify each part by its industry-standard name; this is especially useful when communicating your design ideas to factories or developers. The shoe is composed of numerous parts that are often manufactured independently but still need to work together as a dynamic whole.[1]

Heels and soles, for example, are usually made by experts in completely separate locations. Even the stitching of the upper sections (also known as "**closing**") can be outsourced. A shoe factory is merely a place where all these parts are assembled to produce a shoe. Generally speaking, factories do not produce any components or raw materials.

Following are definitions of the most important shoe components:

The upper is everything on the shoe above the sole. It is made up of pattern pieces that are sewn together. Common upper material is leather (mainly cowhide), but uppers can also be made of other materials such as textiles (e.g., synthetics, fabric, rubber).[2]

The lining is important in keeping the internal parts of the upper in place by supporting it. Common lining materials include **pigskin**, **calfskin**, **kidskin**, and textiles.

A **toe box** helps to maintain the shape and height of the front end of the shoe. It is a piece of **semi-rigid thermoplastic material** that is heat-molded to the shape of the toe area.[3] Finer shoes can have a toe box made of leather.

A heel counter helps maintain the shape of the heel cup area and holds the heel of the foot in place.[4] It is a piece of semi-rigid thermoplastic material. Finer shoes can have a heel counter made of leather.

The sock lining creates a surface that touches the bottom of the foot. It covers either the footbed or the insole, and consists of a piece of leather or fabric. This is where the branding is commonly placed.

The shank acts as a supporting bridge between the heel and the ball of the foot. Attached to the insole board, it is usually a steel strip but can also be made from **nylon**, wood, or even leather.

An insole provides structure and shape to the bottom of the shoe, its main function being a component to which the upper can be attached. It is made up of an insole board and a shank **glued** together. The insole board consists of a cellulose board or a composite material.

An outsole is the bottom part of the shoe that touches the ground. Outsoles can be made from various materials depending on the price and the end use of the shoe. Leather, from bovine animals, is used for higher-end footwear. Materials such as **natural crepe rubber**, **resin rubber**, **polyurethane** (PU), and **vulcanized rubber** are commonly used for soles.

The heel is a raised support of hard material, attached to the sole under the back part of the foot, usually made of hard plastic and covered in leather. **Stacked leather**, wood, or wood covered in leather are also occasionally used in higher-end footwear. The small plastic bottom tip of a woman's shoe-heel is called a **heel cap** or **heel tip**. It is designed to be easily replaced after wear and tear.

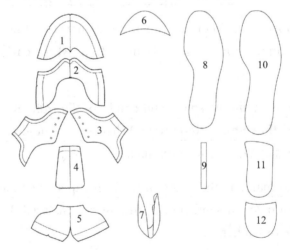

Figure 1   Shoe pattern parts

1 toe cap; 2 vamp upper; 3 quarter; 4 tongue; 5 counter; 6 toe box; 7 heel counter; 8 insole; 9 shank; 10 outsole; 11 half sock lining; 12 heel

重点及难点句

1. The shoe is composed of numerous parts that are often manufactured independently but still need to work together as a dynamic whole.

鞋子由许多独立制作的部件组成，但这些部件仍需要作为一个动态的整体一起发挥作用。

2. Common upper material is leather (mainly cowhide), but uppers can also be made of other materials such as textiles (e.g., synthetics, fabric, rubber).

常见的帮面材料是皮革（主要是牛皮），但帮面也可以用其他材料，如纺织品（如合成材料、织物、橡胶）。

3. A toe box helps to maintain the shape and height of the front end of the shoe. It is a piece of semi-rigid thermoplastic material that is heat-molded to the shape of the toe area.

包头有助于保持鞋前端的形状和高度。它是一块半刚性热塑性材料，利用热成形工艺来形成鞋头形状。

4. A heel counter helps maintain the shape of the heel cup area and holds the heel of the foot in place.

主跟有助于保持足部后跟区域的形状，并使脚处于正确位置。

词　汇

closing：缝帮

pigskin：猪皮

calfskin：小牛皮

kidskin：小山羊皮

toe box：包头

semi-rigid thermoplastic material：半刚性热塑性材料

nylon：尼龙

glue：用胶水将物体黏合，胶水

natural crepe rubber：皱片胶（由天然橡胶加工而成）

resin rubber：树脂橡胶

polyurethane：聚氨酯

vulcanized rubber：硫化橡胶

stacked leather：堆叠皮革

heel cap/heel tip：鞋跟面皮

Lesson 3 Know Your Footbeds

Footbed (or insole or sockliner) is important to the fit, feel, performance and cost of your shoe design. Footbeds come in all shapes, sizes and materials.

Each shoe last will have a "sock allowance" built into the bottom, this creates the space inside the shoe, in can be 4mm, 6mm, or more.[1] You have get this right or your shoe won't fit quite right. A little extra foam in your footbed can be used to fine tune and fitting issues.

The footbed is also key to the lifespan of your shoe. A thin footbed made of too soft, cheap foam can collapse after just a few days leaving the shoe fitting loose and your customer with tired feet.[2] The ability of a foam to survive being placed underfoot is called its compression set or compression resistance.[3] Not all foams are suitable for long-lasting footbed service.

1. Common shoe footbed types

Footbeds basically come in two construction types. Molded or die cut. The molded footbeds are made of **compression molded EVA**, **poured PU (polyurethane foam)**, **latex & cork**, **sponge rubber or PE (polyethylene) foam**. You can add to the molded footbed other features like injection molded stiffeners, support frames, gel pods or airbags.[4]

The molded footbed is a must for performance athletic shoes, the contour will support the foot and hold the foot in place. Hiking, hunting and military boots need molded footbeds. The **strobel shoes** may be thinner as the inside of the shoe will have more contours from the last and molded midsoles. Stiffer board lasted boots should have a molded footbed to provide support and fill up the square corners near the edges.

Die cut does not have to mean cheap. Yes, the most basic shoes will have thin die cut footbeds made of soft EVA foam that will last just a few weeks. High end shoes can have really nice leather-covered die cut footbeds made with multiple layers of high-quality long-lasting PU foam, **neoprene rubber** or **gel sheet**. Fashion and casual shoes can have die cut footbeds. Soccer cleats, the even expensive ones, often have thin die cut footbeds.

2. Footbed cover fabric

Footbeds can come with different cover fabrics or linings. The best fabrics have enough grip to hold your feet in place. Too smooth will not be stable, to grippy will ruin your socks. Running shoes, and hiking boot footbeds need abrasion resistant materials to last for miles and miles. The footbed material must resist crocking (the color transfer by friction or rubbing). Your shoes should not discolor your socks. Leather footbeds are nice, but maybe not the best for athletic shoes.

3. Other footbed features

Footbeds can have multiple densities of foam. They can have perforations or other venting features. **Moisture wicking fabric** covers or bright graphics are a nice touch.[5] Most companies will have their own molds with refined shapes and molded logos. For winter boots or hunting boots you will find insulated footbeds with **heat reflecting coatings**.[6]

重点及难点句

1. Each shoe last will have a "sock allowance" built into the bottom, this creates the space inside the shoe, in can be 4mm, 6mm, or more.

每个鞋楦的底部都有"鞋垫余量",这在鞋子内部创造了4毫米、6毫米甚至更多的空间。

2. A thin footbed made of too soft, cheap foam can collapse after just a few days leaving the shoe fitting loose and your customer with tired feet.

如果鞋垫由廉价的、太软的发泡材料制成,几天后就会塌陷,鞋子就会松垮,消费者的脚就会感到疲劳。

3. The ability of a foam to survive being placed underfoot is called its compression set or compression resistance.

发泡材料被踩在脚下仍能维持原状的能力被称为它的压缩形变或抗压性能。

4. You can add to the molded footbed other features like injection molded stiffeners, support frames, gel pods or airbags.

可以给成形鞋垫增加其他功能，如注塑加强筋、支持框架、凝胶垫或气囊。

5. Moisture wicking fabric covers or bright graphics are a nice touch.

由吸湿织物制成的（鞋垫）表层或者明亮的图案都可赋予其良好的触感。

6. For winter boots or hunting boots you will find insulated footbeds with heat reflecting coatings.

棉靴或猎靴的鞋垫有隔热功能的热反射涂层。

词　汇

compression molded EVA：模压乙烯 – 酸酸乙烯酯共聚物（EVA）

poured PU (polyurethane foam)：浇注 PU（聚氨酯发泡材料）

latex & cork：乳胶和软木

sponge rubber or PE (polyethylene) foam：发泡橡胶或 PE（聚乙烯）发泡材料

strobel shoes：由闯棺工艺制成的鞋

neoprene rubber：氯丁橡胶

gel sheet：凝胶垫

moisture wicking fabric：吸湿面料

heat reflecting coatings：热反射涂层

CHAPTER 2

The Shoe Design

Lesson 1 Specification Drawing

Technical drawings and **line drawings** are essentially the same thing: both are clear drawings of your shoes, but technical drawings would have additional specific information, such as detailed measurements, to explain how the shoes are to be made. Line drawings are drawn simply to illustrate the design of your shoe (without measurements) and can be presented alongside artistic illustrations or in a collection lineup presentation.[1]

The drawings are usually made in one of two ways: by hand or using Adobe Illustrator. **Freehand drawings** are acceptable on an individual designer level, but most larger brands require Illustrator artwork. Nevertheless, a freehand drawing can be almost as good as an Illustrator file. In both cases it is important not to use any artistic license or stylization. This type of drawing is used to show the technical aspect of your drawings and the functionality of your work, so the drawing should be as clear as possible.

You need to show all aspects of the shoe. If only the outside-facing side of the shoe is drawn it is assumed that the inside is the same. The best way to show a technical drawing is from both sides and the top (outside-facing, inside-facing, and overhead views of a shoe), especially if there are details that are not perfectly symmetrical on both sides.[2] Do not forget that you need to think in three dimensions.

Technical drawings are the communication between the designer and the manufacturer. Hence it is vital that the information is as complete as possible and the drawings are at the right scale, which is a good reason for using the actual last as a base for your drawings.[3] In technical line drawings there should not be any stylistic license; draw the lines as they would look on a real shoe. In general, the drawings will be sent via e-mail to the manufacturer to produce a sample. The basic rule is to keep the drawings simple, clear, and labeled with essential information. By this stage it is likely that you, either on your own or with the manufacturer, will have already sourced the last, heel, and the rest of the components, so this is just to tell the manufacturer how to put it all together. It is also a good idea to send a colored version of the line drawing to give a more realistic view of how you expect the final shoe to look.

Computer software enables you to add **textures** and surfaces to your drawings. A hand-drawn starting point, such as a simple line drawing, with computer enhancements will give the best results.

A fully computer-rendered illustration, especially if not done very well, can be cold and spiritless.[4] The computer is, after all, producing the illustration based on your programming. You therefore need to become proficient with the computer software.

To experiment you can start by using a simple color-drop technique, and later move on to leather and other materials. Most computers come with some kind of image-editing program, but Adobe Photoshop is the most popular with professionals. You will find virtually endless ways of working with Photoshop—but to get you started, the following pages show some very basic and quick ways to add color, texture, and **shading** to your product.

Figure 1   Computer—rendered illustruction

重点及难点句

　　1. Line drawings are drawn simply to illustrate the design of your shoe (without

measurements) and can be presented alongside artistic illustrations or in a collection lineup presentation.

线稿只是为了说明鞋子的设计（无尺寸说明），可以与艺术插图或系列线稿一同呈现。

2. The best way to show a technical drawing is from both sides and the top (outside-facing, inside-facing, and overhead views of a shoe), especially if there are details that are not perfectly symmetrical on both sides.

展示技术图纸的最佳方式是从两侧和顶部（鞋的外侧、内侧和俯视图）来展示，尤其是鞋子两侧并非完全对称的设计。

3. Hence it is vital that the information is as complete as possible and the drawings are at the right scale, which is a good reason for using the actual last as a base for your drawings.

因此，信息尽可能完整，图纸的比例尺寸尽可能正确是至关重要的，最好使用真实的鞋楦作为图稿绘制的基础。

4. A fully computer-rendered illustration, especially if not done very well, can be cold and spiritless.

一张完全由计算机渲染的插图，特别是如果做得不是很好，会显得冷漠且缺乏活力。

词 汇

technical drawing：技术图纸

line drawing：线稿

freehand drawing：徒手画，示意图

texture：质感

shading：阴影

Lesson 2   How to Make a Shoe Pattern

Making a shoe pattern or cutting a shoe pattern is not a difficult shoemaking skill. The basic techniques for drawing on a shoe last are simple, and the process is easy, but the skills required to make a beautiful, well-proportioned, mechanically sound, shoe pattern may take years for a shoemaker to perfect. This is the true art of the shoemaker that a computer cannot replace. A skilled pattern cutter is the heart of a shoe factory's development group. Here we will show you how you can cut your own pattern.

1. Shoemaking tools you will need to make a shoe pattern

A shoe last. If you don't have a shoe last you can tape over a shoe or 3D print a last.

A roll of 1/2 inch wide masking tape. This will be used to cover the shoe last. (3/4 inch will do too.)

A sharp X-Acto knife or other hobby knives.

Pens and pencils.

A small flexible steel ruler.

Bristol paper (any stiff paper will do).

2. Steps to make your own shoe pattern

(1) Tape up the shoe last

To make a new shoe pattern step-by-step you will first need to "tape" the last so you can "pull the shell" off the shoe pattern. The tape should be layered in two different directions so the pattern stays together when you are ready to peel it off the last.

You will start on the lateral or outside of the shoe last. First, starting at the top, lay the tape lengthwise down the lateral side of the last. Next, run a strip of tape down the center of the last from the top of the instep down to the toe. Do the same on the heel of the last. Finally, run layers across the last from the bottom edge up the side of the last. Make this extra smooth, this will be your drawing surface when you start marking your shoe pattern.

Make sure the tape wraps around the bottom edge of the last. You will need this edge, as it will become the bottom edge of your shoe pattern. If your shoe design is symmetrical, then half the last is okay. If the medial and lateral sides of your shoe design are different you will need to tape the entire last. The process is the same for both sides.

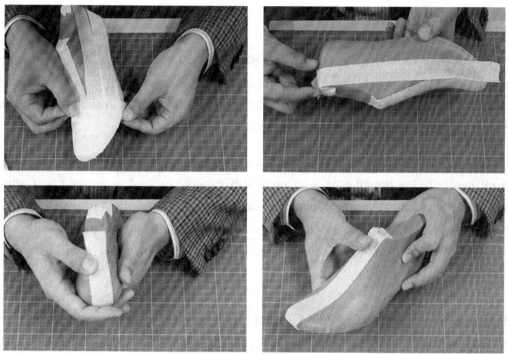

Figure 1   Tape up the shoe last

(2) Mark the shoe pattern on the last

Once the shoe last is covered with the tape, it is time to start marking the shoe pattern. With the steel ruler, mark the centerline of the last from the instep to the toe tip and down the heel. With the centerlines marked it is time to start drawing the pattern on the last.

(3) Drawing the shoe pattern

It's fun to add a little color or paste on a logo. You can start to get a feel for what the design will look like in 3D. Go ahead and iterate upon your design, this is your time to be creative. If you don't like your line, grab some tape, cover it up, and try again.

Make a quick check of the top line, **collar**, and **heel notch** heights.[1]

The design does not have to be perfect now but keep an eye out for any major flaws. This is a good time to show it to your developer. Are the pattern parts wasting material? Are there any overlapping issues? Too many layers overlapping in the flex zone will cause problems.

(4) Flatten the shoe pattern

After you have your shoe design down on the tape it is time to peel off the tape. Using a sharp blade, cut the tape down the center lines of the toe and heel. Next, cut the tape along the bottom edge of the last. Now, start peeling off the tape. If the tape was laid down correctly in overlapping layers the pattern will pull off.

Here is the pattern free of the last. You may find that it does not lay flat, don't worry. Now, carefully lay the tape on a sheet of stiff paperboard and press your new shoe pattern flat. Starting at the top eyelet position and collar line, work your way down the middle then out to each end. As you move to the heel and toe, the 3D pattern will need some relief cuts to "**spring**" the pattern.[2] Add a few cuts and the pattern will flatten.

(5) Cut the shoe pattern

Once the tape is peeled from the last and flattened you have the shoe pattern! The pattern with all the parts together is called the "shell" pattern. With the pattern attached to some paper, it's time to "trim out" the pattern. Now the shoe pattern can be redrawn in a computer and cut out of heavy paper. The **pattern cutter** will add the overlaps and alignment marks.

As a shoe designer, you should always ask for the flat pattern for a new shoe design. When you make corrections for the pattern maker, it's easier to draw on the flat pattern than to draw on the sample. It is also much faster to scan the corrections and email them to the factory rather than deliver the shoe back.

Here is the finished flat pattern. Scanned and with overlaps added, this pattern is ready for the sample room.[3]

重点及难点句

1. Make a quick check of the top line, collar, and heel notch heights.

快速检查一下背中线、鞋口以及后帮高度。

2. As you move to the heel and toe, the 3D pattern will need some relief cuts to "spring" the pattern.

当展平到后跟和鞋头部位时，需要在三维样板上打一些剪口进行跷度处理。

3. Scanned and with overlaps added, this pattern is ready for the sample room.

扫描并添加压茬部分，此样板即可在样品室中使用。

词 汇

bristol paper：高级绘图纸

collar：（运动鞋等的海绵）护口，装饰性沿口

heel notch：后帮上口（V 字型口）

spring：（前后）跷，（鞋样板）升跷，起跷，样跷

pattern cutter：样板切割机

Lesson 3   Designer Competencies

Footwear designers refer to those who design various types of footwear products according to the human foot shape, movement mechanism and aesthetic principles, combined with the nature of shoe-making materials and manufacturing processes.[1] They need to carry out market research, collect and organize popular elements and grasp popular trends; be familiar with color matching and materials and fabrics; be able to express design concepts through line drawings, 2-dimensional drawings and 3-dimensional drawings; understand production processes. The following is a detailed description of shoe and boot designer competencies:

Get inspiration

For some, finding inspiration is easy, for others less so. But where does one find inspiration? Inspiration can come from anything that surrounds you. Ideally, inspiration should come from personal observation, and should not be borrowed from fashion magazines or other designers' work. The theme for your collection will come from your inspiration.

Research ability

Investigation is the actual gathering of information about the theme that emerged as a result of the inspiration stage. This is also when your **sketchbook** really starts to develop. Investigative action is broken down into two types of research: primary and secondary. Primary research is when you personally gather the information or replicate it. Secondary research is when someone else has done the work, such as if you are using existing research that appears on the Internet or in books and magazines.[2]

Materials and colors

To help you communicate about color with others working in the industry, internationally recognized color-matching systems based on codes are used, such as **Pantone**.[3] Using such systems is essential, especially when communicating digitally. You might have a swatch and a beautifully illustrated shoe, but a scan of this viewed on another person's computer could look totally different.

Learn to use material boards to assist in selection and color matching. The materials board should show textures and other surface references that would not display well on a flat color board. Using the color palette from your color board as a base for your materials board, print a Pantone-coded sheet of your palette on 11×17 in. Find materials that are the closest to your colors. If you do not have access to real materials, you can use a material that is similar to get your point across (a papery finished leather could be replaced by lightly wrinkled paper). The Internet and hardware, fabric, craft, and thrift stores are great places for finding excellent material references.

Realize the design idea

Have the ability to draw shoes, and be able to clearly express design ideas by using 2D or 3D drawing software, including but not limited to the design renderings and structural drawings of shoes. Technical drawings are the communication between the designer and the manufacturer. Hence it is vital that the information is as complete as possible and the drawings are at the right scale.

Understand process technology

A career in shoemaking involves working in a niche market and requires a multitude of very specialized skills. The best path is to take a technical course. Be able to understand the principle of shoemaking technology and practical operation process. Be able to know shoemaking machinery.

重点及难点句

1. Footwear designers refer to those who design various types of footwear products according to the human foot shape, movement mechanism and aesthetic principles, combined with the nature of shoe-making materials and manufacturing processes.

鞋履设计师是指根据人的足形、运动机制和美学原理，结合制鞋材料的性质和制造工艺，设计各种鞋类产品的人。

2. Primary research is when you personally gather the information or replicate it.

Secondary research is when someone else has done the work, such as if you are using existing research that appears on the Internet or in books and magazines.

一手调研是自己亲自收集或复制这些信息；二手调研指别人已经做了这项工作，比如你使用的是互联网上或书籍和杂志上的已有研究。

3. To help you communicate about color with others working in the industry, internationally recognized color-matching systems based on codes are used, such as Pantone.

为了帮助您与业内工作人员交流色彩方面的信息，应使用国际公认的基于代码的配色系统，如潘通色卡。

词 汇

sketchbook：速写本，素描册

Pantone：潘通，潘通色彩系统

CHAPTER 3

Customization of Shoes

Lesson 1　Mass Customization

The traditional system of premanufacturing and traditional retail shows limitations: although the product variety and availability seem limitless, customers leave stores without the shoes they were hoping to find-color, size, style, or fit may not be available at the moment in time the customer is looking for it.

A market research on customization had the following key findings:

Fit is a critical buying factor: 70% are interested in personalized fit, accepting a higher price.

Customers are unhappy about the availability in stores: 84% have experienced that a shoe they intended to buy was not available, and 60% experience this often.

Buying shoes is very emotional: 35% buy shoes just for emotional reasons and another 35% buy to replace other shoes.

With markets long being **saturated**, customers becoming more and more demanding, and technology and organizational methods evolving a new field for "mass customized" footwear is opening.[1] Mass customization and customer co-design options offer important and valuable options for differentiation.

1. Style customization in shoes

Style customization in shoes aims at satisfying customers' diverse demands on shoe aesthetics and self-identity reflected by the shoes. In style customization, based on the standard lasts and sizes, the customer can choose from diverse style options, including colors, fabrics, leather, and other accessories, within the constraints set by the provider. It offers customers the freedom to configure shoes by selecting their favorite options from a list of possible variants based on the standard of shoe design. This strategy has been widely adopted by many large shoe producers because it requires less effort from both producers and customers. Although style customization involves high costs in design sales, it has little effect on manufacturing and production costs.

2. Technical approaches in shoe customization

From the customer's perspective, the whole shoe customization process can be broken down into four steps: search for favorite "basic" shoes, co-design and configure the shoes, make an order to purchase the customized shoes, and receive the shoes delivered to home soon.

To support the customization process, the producer should reconsider the whole value chain from the front end—including product portfolio management, customer interface design, and product **configurator**—to the back end including flexible manufacturing and supply chain management,[2] as shown in Figure 1.

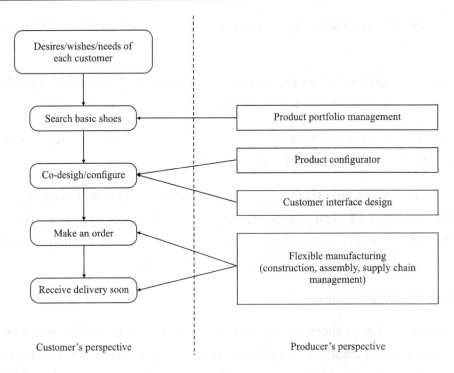

Figure 1   Technical approaches in shoe customization

重点及难点句

1. With markets long being saturated, customers becoming more and more demanding, and technology and organizational methods evolving a new field for "mass customized" footwear is opening.

随着市场长期处于饱和状态，客户要求越来越高，技术和组织方式的发展开辟了"大规模定制"鞋类的新领域。

2. To support the customization process, the producer should reconsider the whole value chain from the front end—including product portfolio management, customer interface design, and product configurator—to the back end including flexible manufacturing and supply chain management.

为支持定制过程，生产商应该重新考虑从前端到后端的整个价值链，前端包括产品组合管理、客户界面设计和产品配置，后端包括柔性制造和供应链管理。

词 汇

saturated：饱和的
configurator：配置，配置程序

Lesson 2   Reasons to Enter Mass Customization

The generation growing up with smart phones, the Internet, and online games are used to configure their online environments and tools. They will expect to be able to configure traditional products as well. They will also have the skills to do so. With the technology of product **configurators** getting more refined and affordable, consumers will want more say, and the "old world" of footwear manufacturing will have to find a way to manufacture based on customer orders.

In today's highly competitive business environment, activities for serving customers have to be performed both efficiently and effectively—they have to be organized around a customer-central supply and demand chain.[1] Since the early 1990s, mass customization has emerged as an idea for achieving precisely this objective.

As a business paradigm, mass **customization** provides an attractive business proposition to add value by directly addressing customer needs and in the meantime, utilizing resources efficiently without incurring excessive cost.

As we wear shoes every day, most of us have quite a precise idea of a perfect shoe. But in reality, there is often a trade-off in terms of fit, form, and function. As a result, the footwear sector has been among the industries that have embraced mass customization quite early.

Regardless of different product categories, price points, and fulfillment systems, they all have turned customers' heterogeneous needs into an opportunity to create value, rather than regarding heterogeneity as a problem that has to be minimized, challenging the "one size fits all" assumption of traditional mass production.[2]

The global shoe market is characterized by a high level of rivalry between the existing suppliers. Acquisitions such as the one of Reebok by Adidas AG in 2005 or Umbro by Nike in 2007 lead to permanent movements of market shares. Furthermore, low entrance barriers allow new, mostly specialized suppliers to enter the market, even if they are only regional competitors of the global players.[3] Most companies have outsourced their production facilities to low-wage countries in the Far East. Simultaneously, shortened innovation cycles have led to higher costs in research and development as well as less time for product and quality testing before a product enters the market. From the consumer's perspective, the Internet has created an environment where customers are confronted with a huge bandwidth of products as well as more information to compare these[4] (specialized category retailers like **Zalando**, an online footwear shop, are among the most successful examples of e-commerce). In most cases, a change to another brand evokes no costs. As a result, quality aspects as well as a psychological brand commitment and a perfect fit to customers' needs are becoming key enablers of a strategy in the shoe industry.

重点及难点句

1. In today's highly competitive business environment, activities for serving customers have to be performed both efficiently and effectively—they have to be organized around a customer-central supply and demand chain.

在当今高度竞争的商业环境中，服务客户的活动必须有效且高效地执行——它们必须围绕以客户为中心的供需链进行组织。

2. Regardless of different product categories, price points, and fulfillment systems, they all have turned customers' heterogeneous needs into an opportunity to create value, rather than regarding heterogeneity as a problem that has to be minimized, challenging the "one size fits all" assumption of traditional mass production.

无论不同的产品类别、价格点和执行体系如何，它们都将客户的异质需求转化为创造价值的机会，而不是将异质视为一个必须最小化的问题。这是对传统大规模生产"一刀切"假设的挑战。

3. Furthermore, low entrance barriers allow new, mostly specialized suppliers to enter the market, even if they are only regional competitors of the global players.

此外，较低的准入壁垒允许新的、主要是专营供应商进入市场，即使他们只是全球参与者的区域竞争对手。

4. From the consumer's perspective, the Internet has created an environment where customers are confronted with a huge bandwidth of products as well as more

information to compare these.

从消费者的角度来看，互联网创造了一个环境，在这个环境中，消费者面对着巨大的产品范围以及更多的信息来比较这些产品。

词 汇

configurator：配置，配置程序

customization：定制，专用化

Zalando：总部位于德国柏林的大型网络电子商城

Lesson 3 Footwear Customization for Manufacturing Sustainability

1. Mass customization

Mass customization refers to the customer co-design process of products and services, which meets the needs of each individual customer with regard to certain product features. All operations are performed within a fixed solution space, characterized by stable but still flexible and responsive processes. As a result, the costs associated with customization allow for a price level that does not imply a switch in an upper market segment.[1]

ITEM 1: Customer Co-design: Customers are integrated into value creation by defining and configuring an individual solution.[2]

ITEM 2: Meeting the Needs of Each Individual Customer: To better express the level of customization offered, we highlight three dimensions: fit, style, or functionality. In the case of footwear, this means to measure the two feet in three dimensions and extract the necessary information to choose the best fitting last or even to make the personalized one.

ITEM 3: Stable Solution Space: Setting the solution space becomes one of the foremost competitive challenges of a mass customization company.

ITEM 4: Adequate Price: Mass customized goods are thus targeting the same market segment that was purchasing the standard goods before but with the adequate price increase.

2. Sustainability

The concept of sustainability, in the way we understand the term now, first **hit the ground** in 1987, with the Brundtland Report, defining it as "to meet the needs of the present generation without compromising the ability of future generations to meet their own needs".[3] Later, as the concept gained popularity and momentum, hundreds of definitions have been proposed, in **academic debates** and business arenas, referring to a more humane, more ethical, more green, and more transparent way of doing business.

In the jungle of definitions, and to be able to point out its link with mass customization, we try hereinafter to set some **cornerstones**, exploring the three sustainability **pillars**: economical, environmental, and social.

The environmental dimension of sustainability denominates those aspects that mainly describe environmental performances, in order to minimize the use of hazardous or toxic substances, resources, and energy. These aspects can be summarized as follows: emissions, use of resources, and waste.

Economical aspects can be summarized as follows: efficiency, investment in technologies and competences, and **supply chain risk**.

Social dimension of sustainability usually encompasses aspects such as social responsibility, health and safety, the "polluter pays" principle, and reporting to the **stakeholders**.[4] Social aspects can be summarized as follows: working conditions and workforce, product responsibility, and local community benefits.

重点及难点句

1. All operations are performed within a fixed solution space, characterized by stable but still flexible and responsive processes. As a result, the costs associated with customization allow for a price level that does not imply a switch in an upper market segment.

（大规模定制）所有操作都在一个有固定的解决方案的空间内执行，其特点是过程稳定但仍然灵活且响应迅速。因此，与定制相关的成本允许有一个价格水平（浮动），但并不意味着转向了高端市场。

2. Customers are integrated into value creation by defining and configuring an individual solution.

通过定义和配置个性化解决方案，将客户融入价值创造中。

3. The concept of sustainability, in the way we understand the term now, first hit the ground in 1987, with the Brundtland Report, defining it as "to meet the needs of the present generation without compromising the ability of future generations to meet their own needs."

按照我们现在对该术语的理解，可持续性发展的概念首次出现在 1987 年的《布伦特兰报告》中，其定义为"在不损害后代人满足他们自己需求能力的情况下满足当代人的需求。"

4. Social dimension of sustainability usually encompasses aspects such as social responsibility, health and safety, the "polluter pays" principle, and reporting to the stakeholders.

可持续性的社会维度通常包括社会责任、健康和安全、"污染者付费"原则以及向利益相关者报告等方面。

词 汇

mass customization：大规模定制

hit the ground：落地（指概念被人们知晓）

academic debate：学术辩论会

cornerstone：基石

pillar：支柱

supply chain risk：供应链风险

stakeholder：利益相关者

PART III
Shoemaking Technology

CHAPTER 1

Manufacturing Process of Shoes

Lesson 1   Shoemaking Process

A footwear company has mainly four departments in which a progressive route is followed for producing finished shoes. These are—**clicking or cutting department, closing or machining department, lasting and making department, finishing department and shoe room**.

1. Clicking or cutting department

In this department, the top part of the shoe or the "upper" is made. Upper clicking means cutting all upper components of various materials out as accurately and economically as possible, at the required quality standard through various means and processes either manually or mechanically.[1] Leather may have various defects on the surface such as **barbed wire** scratches which need to be avoided so that they are not used for uppers.

Owing to the high cost of material, particularly leather, and considering the fact that it will affect most of the subsequent manufacture operations and will, above all, result in the quality standard of the finished product, clicking can be considered as one of the most important steps in the footwear manufacturing process.

2. Closing or machining department

Here the component pieces are sewn together by highly skilled workers. The work is divided into stages. In the early stages, the pieces are sewn together on the **flat machine**. In the later stages, when the upper is no longer flat and has become **three-dimensional**, the machine called **post machine** is used. The sewing surface of the machine is elevated on a **post** to enable the operative to sew the three-dimensional upper.[2] Various **edge treatments** are also done onto the leather to give an attractive look to the finished upper. At this stage only, the **eyelets** are also inserted in order to accommodate the **laces** in the finished shoes.

3. Lasting and making department

The completed uppers are molded into the shape of a foot with the help of a "last". Last is a plastic shape that simulates the foot shape. It is later removed from the finished shoe to be used further in making other shoes. Firstly, an **insole** is attached to the bottom of the last. It is only a temporary attachment. Sometimes, mostly when **welted shoes** are manufactured, the insole has a rib attached to its under edge. The upper is stretched and molded over the last and attached to the insole rib. After the procedure is completed, a "lasted shoe" is obtained. Now, the welt—a strip of leather or plastic—is sewn onto the shoe through the **rib**. The upper and all the surplus materials are **trimmed** off the **seam**.[3] The sole is then attached to the welt and both are stitched together. The heel is then attached which completes the making of the shoe.

4. Finishing department and the shoe room

The finishing of a shoe depends on the material used for making it. If made of leather, the sole edge and heel are trimmed and buffed to give a smooth finish. To give them an attractive finish and to ensure that the edge is waterproof, they are **stained**, polished and **waxed**. The bottom of the sole is often lightly **buffed**, stained and polished and different types of patterns are marked on the surface to give it a craft-finished look.[4] A "finish shoe" has now been made.

For shoe room operation, an internal sock (or **shoe tree**) is fitted into the shoe which can be of any length—full, half or quarter.[5] They usually have the manufacturers' details or a brand name wherever applicable. Depending on the material used for the uppers, they are then cleaned, polished and **spread**. Laces and any tags that might have to be attached to the shoes, such as shoe **care instructions**, are also attached. The shoes, at last, get packaged in boxes.

重点及难点句

1. Upper clicking means cutting all upper components of various materials out

as accurately and economically as possible, at the required quality standard through various means and processes either manually or mechanically.

帮面切割是指通过各种手动或机械的方式，按照所需的质量标准，尽可能准确和经济地切割出不同材料的所有帮面部件。

2. The sewing surface of the machine is elevated on a post to enable the operative to sew the three-dimensional upper.

机器的缝纫面被抬高到一个立柱上，使操作者能够缝制三维的鞋面。

3. The upper and all the surplus materials are trimmed off the seam.

鞋脚和所有多余的材料会从接缝处修剪掉。

4. The bottom of the sole is often lightly buffed, stained and polished and different types of patterns are marked on the surface to give it a craft-finished look.

鞋底面通常经过轻微打磨、染色和抛光，表面压印不同类型的图案，使其具有精致的外观。

5. For shoe room operation, an internal sock (or shoe tree) is fitted into the shoe which can be of any length—full, half or quarter.

后整饰时在鞋腔内装入鞋撑，这个鞋撑可以是全长、半长或四分之一长度。

词 汇

clicking or cutting department：裁断车间

closing or machining department：机缝车间

lasting and making department：绷帮和制底车间

finishing department and shoe room：成品车间

barbed wire：铁丝网	flat machine：平板缝纫机
three-dimensional：三维	post machine：高桩柱缝纫机
post：缝纫机柱，架	edge treatment：（皮料）边口处理
eyelet：鞋眼	laces：鞋带
insole：鞋垫	welted shoe：沿条鞋
rib：内底埋条	trim：修边
seam：接合（处，线，缝）	stain：着色

wax：打蜡　　　　　　　　　　　buff：砂磨

shoe tree：鞋撑　　　　　　　　　spread：遮盖（缺陷）

care instruction：保养说明，使用说明

Lesson 2   Knit Shoe Construction

The demand for leather shoes is declining considerably due to environmental consciousness, resulting in a corresponding rise in demand for textileshoes. Athletic Footwear (Sport Shoes) accounts for about 36 percent of all sales of footwear worldwide. The use of knit technology in upper making, especially in the case of athletic footwear, could transform the entire traditional shoemaking process.

A one-piece upper design is produced by a CNC **knitting** machine and then assembled with the tongue, lining materials, and reinforcements.[1] Because knit shoe construction can radically simplify shoe construction, the technique has spread from **high-end** to lower-priced shoes in just a few years. You can find running shoes made by many famous brands, and even Chinese local market **casual shoes** are now being made with 4D knitting technology.

In the major shoemaking areas of China, the **CNC** knitting machine is becoming a common sight. With rising labor costs in major shoe manufacturing countries like China, the shoe knitting machine, which can knit one upper in 10 minutes and **runs 24-7**, is seen as a labor-saving device.[2] Knit shoe construction has also opened the door to new, more **sustainable** production with less waste.

Knit shoe construction

To construct the one-piece upper, the knitting machine is loaded with **polyester**, nylon, or **spandex fibers**. The newest knitting machines can handle a mix of fibers and up to 10 colors at one time. The machine can be programmed to knit one upper at a time or 3 uppers with a maximum width of 90cm.[3] Depending on the programming and fibers selected, the upper can be thin and stretchable or thick and stretch-resistant. The design opportunities are nearly infinite with fiber options, color choices, knit densities, and opening configurations.[4]

While these shoes look like they are entirely constructed with the knit material, the internal **linings**, **reinforcements**, and padding are exactly what you will find on a **conventionally** cut and stitched shoe.

Of course, the knit upper shoe is now evolving in new directions. The naturally breathable designs combined with stretchable and non-stretchable fibers are creating new opportunities for footwear designers. Shoe uppers can be made in a single knitted layer with all the functionalities needed and appropriate. A tighter weave can be built to provide more **arch support** to the foot by changing the knit density in different areas and a thinner, **breathable** weave to provide more air flow.[5]

Knitting Cost

The cost depends on the number of colors and fiber types. A single-color polyester fiber design may cost $2.50, while a multi-color, **polyester spandex** combination can cost $7.50 per upper.

MOQ (Minimum Order Quantity) is around 1000 pairs, with a setup fee per design of $300. Sample development is fast—just 7 to 10 days, this may take longer if custom color fibers are required. Once the design is confirmed the automated production can run around the clock. 1000 parts can be done in 7 to 10 days, less depending on how many knitting machines are deployed.

Shoe knitting is an independent process, like 3D printing, which saves time, money and energy. This commonly used 3D knitting technique will improve the productivity of the industry by reducing material costs, reducing labour-intensive costs and keeping time running quickly.

With the implementation of this technology, 80% less waste can be generated. Resulted factors such as global warming potential, **eco-toxicity** and non-**carcinogenic** effects on human health indicate that the development of knitted uppers has a lower environmental impact than conventional running shoes.[6]

重点及难点句

1. A one-piece upper design is produced by a CNC knitting machine and then assembled with the tongue, lining materials, and reinforcements.

一片式鞋面由数控针织机生产，然后与鞋舌、衬里材料和加强件组装在一起。

2. With rising labor costs in major shoe manufacturing countries like China, the shoe knitting machine, which can knit one upper in 10 minutes and runs 24–7, is seen

as a labor-saving device.

随着像中国这样的主要制鞋国家劳动力成本的上升，可以在 10 分钟内编织一个鞋面并按照全天候模式（每天 24 小时、每周 7 天）运行的制鞋机被视为一种节省劳动力的设备。

3. The machine can be programmed to knit one upper at a time or 3 uppers with a maximum width of 90cm.

该机器可编程为一次编织 1 个鞋面或 3 个最大宽度为 90 厘米的鞋面。

4. The design opportunities are nearly infinite with fiber options, color choices, knit densities, and opening configurations.

纤维、颜色、编织密度和网眼配置的多种选择使设计具有无限可能。

5. A tighter weave can be built to provide more arch support to the foot by changing the knit density in different areas and a thinner, breathable weave to provide more air flow.

更紧密的编织可以通过改变不同区域的编织密度来提供更多的足弓支撑，而更薄、透气的编织可以提供更多的空气流动。

6. Resulted factors such as global warming potential, edification, eco-toxicity and non-carcinogenic effects on human health indicate that the development of knitted uppers has a lower environmental impact than conventional running shoes.

由此产生的全球变暖潜势、生态毒性和对人体健康的非致癌效应等因素表明，发展针织鞋面对环境的影响低于传统跑鞋。

词 汇

knit：编织   high-end：高端的

casual shoes：休闲鞋

CNC（Computer Numerical Control）：电脑数控

run 24-7：全天候运行的   sustainable：可持续的

polyester：聚酯，涤纶   spandex fiber：氨纶纤维

lining：衬里

reinforcement：定型材料，补强衬料   conventionally：传统地

arch support：足弓支撑   breathable：透气的

polyester spandex：聚酯氨纶

MOQ (Minimum Order Quantity)：最小订货量

with the implementation of：随着……的实施   eco-toxicity：生态毒性

carcinogenic：致癌的

Lesson 3   RF Welding and Compression Molding for Shoes

Radio frequency welding (RF welding) and **compression molding** are widely used to make modern athletic shoes. Both processes are used to create logos, design details, and even the shoe structure.

Compression molding and RF welding both use heat and pressure to remold shoe materials.[1] Welding can create unique design effects and save labor costs.

1. Welding materials

Man-made and natural materials can both be used but they require different tooling and RF welding techniques. Man-made materials, such as **mesh**, PU, and **foam**, can **melt**, mold, **fuse**, and stretch. Natural materials, such as leather and **canvas**, have some stretchability but will not remelt or fuse without the use of a **synthetic backing material**.[2]

The factory will use EVA, PU, PE, and PVC foams to make the backing layers. These foams all accept the compression and with the right head and pressure, they will not **rebound** after RF welding.[3]

2. RF welding tool

Logo welding equipment is usually Computer Numerical Control (CNC) cut from a soft, **conductive**, and **rust-free metal**, usually **brass**.[4] Larger pressing tools can be cut by CNC from flat aluminum sheets. Pressing tools that require a flat surface to press against can be made one-sided. The one-sided welding tool is limited in the dimensions it can create. A two-sided or "matched welding tool" can create 3D **pockets** and larger features. After molding, these pockets can be filled with foam or silicon.[5]

3. Welding equipment

Two classes of equipment are used to weld and **emboss**. The radio frequency or RF welder uses focused energy to **locally** heat the shoe materials and the machine comes in many sizes. A small, **tabletop** unit is great for logos, while a square meter sized machine is best for clothing and entire shoes.

重点及难点句

1. Compression molding and RF welding both use heat and pressure to remold shoe materials.

模压成型和射频焊接都使用热能和压力来重塑制鞋材料。

2. Natural materials, such as leather and canvas, have some stretchability but will not remelt or fuse without the use of a synthetic backing material.

天然材料，如皮革和帆布，具有一定的延伸性，但如果不使用合成的背衬材料，就不会重新熔化或熔合。

3. These foams all accept the compression and with the right head and pressure, they will not rebound after RF welding.

使用合适的压头和压力时这些泡沫都可被压缩，射频焊接后不会反弹。

4. Logo welding equipment is usually Computer Numerical Control (CNC) cut from a soft, conductive, and rust-free metal, usually brass.

标志的焊接模具通常是由计算机数控切割机从导电和无锈的软质金属（通常是黄铜）上切割而成。

5. A two-sided or "matched welding tool" can create 3D pockets and larger features. After molding, these pockets can be filled with foam or silicon.

双面或"匹配焊接工具"可创建 3D 空腔和更明显的特征。模压成型后，可在这些空腔中填充泡棉或硅胶。

词　汇

radio frequency welding (RF welding)：射频焊接

compression molding：模压　　　　　mesh：网状材料

foam：泡沫，海绵　　　　　　　　　melt：熔化，塑化

fuse：熔合　　　　　　　　　　　　canvas：帆布

synthetic：合成的

rebound：回弹

rust-free metal：防锈金属

pocket：空腔，夹袋

locally：局部的

backing：衬里，背衬

conductive：导电的

brass：黄铜

emboss：压花，装饰

tabletop：桌面，台面

CHAPTER 2

Upper Making Process

Lesson 1   Pattern Lay Out

1. Homogenous materials (woven)

The **pattern lay out** will vary in relation with the kind of material used. For **homogeneous materials** the basic rules should be followed: **plain materials**— only the **stretch/strength directions** have to be considered; **patterned materials**— **pattern lines/design orientation and position** plus stretch/strength directions have to be considered; **woven materials**—pattern lines/design orientation and position plus stretch/strength directions have to be considered.

In general, woven materials present these possibilities: along the warp, almost no stretch; across the warp, some stretch; in **bias**, much stretch.[1]

When cutting fabrics for **vamp lining**, we shall consider only the stretch and strength directions. During the **lasting** process, both the **upper** and the **lining** are exposed together to a very strong pulling action. The upper material will expand relatively and, if the lining cannot expand in the same proportion it will break. Therefore, these vamp linings have to be cut in bias.[2]

A definite **patterns' lay out system** has to be devised for each product and material.

2. Heterogeneous materials (leather)

Principles

We have to consider the considerable variations of the leather material and, as a consequence, the corresponding difference in quality level for what concerns texture, resistance, stretch and strength, color and grain plus the accidental defects affecting many **hides and skins**.[3] Because of that, it is not possible to obtain leather uppers of a uniform quality standard. Therefore we have to accept the fact that some components would be better and nicer than others. It even goes to the extent that a component may show some difference between one area and another.[4]

These lead to a kind of **hierarchy of the components** and areas, and to the definition of some principles in order to obtain the best component quality standard for every heterogeneous materials.

These principles are:

Cut **tight to toe** as far as strength direction is concerned;

Use the best part of the leather for the front part of the shoe;

Give priority to the **outside side** over the **inside side** of the shoe;

Place the **top side** of the components towards the **heart of the skin** and the **lower side (lasting margin)** towards the **edge of the skin**;[5]

Cut from **tail** to **neck**, cutting first **the noblest components**;

Cut from **side to side** if it is a **full skin**;

Cut from the **backbone** side to the **belly** if it is a **side (half skin)**;

Do not place any visible defect in a very visible part of the shoe;

Do not accept a bad defect even if not visible (deep cut on the flesh side) in any components;

Put acceptable defects in the lasting margins and the **underlays** for economy purposes.

Try to cut as horizontally as possible and avoid wedge-like cutting, as this is very prejudicial to material economy.[6]

Although it is not possible for the man-made materials to devise the definite pattern lay out system, the ideas have to be applied as much as possible .

Note: Correct patterns lay out is always a combination, which permits to obtain the maximum number of components of the required quality standard. It always involves these two notions: quality and material economy.

重点及难点句

1. In general, woven materials present these possibilities: along the warp, almost no stretch; across the warp, some stretch; in bias, much stretch.

一般来说，纺织材料具有这些可能性：沿经纱方向几乎没有延伸性；垂直于经纱方向有一些延伸性；与经纱成一定角度的方向则有很大的延伸性。

2. The upper material will expand relatively and, if the lining cannot expand in the same proportion it will break. Therefore, these vamp linings have to be cut in bias.

帮面材料会被拉伸，如果衬里不能以相同的比例延伸就会破裂。因此，这些鞋面衬里必须斜向裁剪。

3. We have to consider the considerable variations of the leather material and, as a consequence, the corresponding difference in quality level for what concerns texture, resistance, stretch and strength, color and grain plus the accidental defects affecting many hides and skins.

我们必须考虑皮革材料的巨大差异，以及由此在质量层面产生的相应差异，这些差异涉及质地、牢度、延伸和强度、颜色和纹理以及对许多皮革面积有影响的意外瑕疵。

4. It even goes to the extent that a component may show some difference between one area and another.

甚至同一部件的不同区域都可能有一定的差异。

5. Place the top side of the components towards the heart of the skin and the lower side (lasting margin) towards the edge of the skin.

使部件的上部靠近皮心部位，下部（绷帮余量处）靠近皮革边缘部位。

6. Try to cut as horizontally as possible and avoid wedge-like cutting, as this is very prejudicial to material economy.

尽量进行水平（排料）切割，避免楔形切割边，因为这极不利于节省材料。

词　汇

pattern lay out：排料（在材料上排布下料样板）

homogeneous material：均质材料

plain material：单色的 / 无图案花纹的光面材料

stretch / strength directions：拉伸 / 强度方向

patterned material：有图案、花纹的材料

pattern lines：图案线条

design orientation and position：图样方向和位置

woven material：纺织材料　　　　　　bias：倾斜，与……成一定的角度

vamp lining：前帮衬里　　　　　　　lasting：绷帮

upper：帮面　　　　　　　　　　　　lining：衬里

patterns' lay out system：排料方案　heterogeneous material：非均匀材料

hide and skin：皮革（hide 一般指面积较大的，如牛皮；skin 一般指面积较小的，如羊皮）

hierarchy of the components：部件的主次之分

tight to toe：紧靠边地　　　　　　　outside side：外怀（部件）

inside side：里怀（部件）　　　　　　top side：（部件的）上口边

heart of the skin：皮心（部位）　　　lower side：（部件的）下口边

lasting margin：绷帮余量　　　　　　edge of the skin：皮革边缘（部位）

tail：尾部　　　　　　　　　　　　　neck：颈部

the noblest component：最主要的部件

side to side：（沿背脊线）对称地　　　full skin：整张革

backbone：背脊线　　　　　　　　　belly：边腹部

side (half skin)：半月革　　　　　　　underlay：可被遮盖的部位

Lesson 2 Upper Cutting and Closing Operations

Upper cutting is the result of two activities of different natures: **cutting operation** and **material exploitation/utilization**. Material exploitation/utilization is a matter of knowledge and it is the most important aspect of upper cutting for what concerns quality and material economy.[1]

For homogeneous materials the knowledge required is little because, when the best exploitation method has been determined for a specific product and a given material, all that has to be done is to follow the method.[2] In many cases homogeneous materials can be cut in multiple layers.

For leather materials, the problem is completely different because of the numerous **irregularities** inherent to its nature. No definite exploitation method can be determined, only general principles, which have to be applied to, as much as each piece of material permits.[3] This is why the **cutter or clicker of leather** plays an important role in the actual manufacturing cost of the product and, as a result, in the final **profit or loss** experienced by the firm.

Upper making/closing is a succession of preparation and sewing operations of which sequential order will depend on the model of the upper or the technical process.[4]

Upper preparation will be considered as covering all non-sewing operations. The main operations and their purpose are:

· **Splitting** (sometimes considered as the upper cutting operation): to even leather thickness; to reduce leather thickness on the whole area; to reduce leather thickness on the determined area only (this is a **skiving** operation).

· **Stitch marking**: to indicate correct upper components positions; **decoration seams** and **decoration perforations**.

· **Identification marking/stamping**: to print **size-fitting series No**. etc. for manufacturing purposes and merchandize identification.

· Skiving: to reduce the components' edge thickness for edge treatment and for component assembling. The top piece usually is **thinned** to half thickness while the bottom is "skived to zero".[5]

· **Gimping**: for components edge decoration purposes.

· **Punching**: for **fitting fastening** (eyelet, buckles, etc.) and for **hand stitching** (moccasin).

· **Perforating**: for upper decoration purposes.

· **Cementing** (or adhesive applying): on edges or the whole component area using different types of **adhesive** and different applying methods.

· Edge treatment: to decorate; to give a better finish and to reinforce.

· **Edge folding** (turning over the edge): without or with **reinforcement**.

· **Edging** (burning the edge): also called (ironing/setting/searing/burnishing).

· Dying: **raw edge**

· **Binding**: mainly consists in applying adhesive and turning over the binding materials.

· **Edge beating**: to reduce the bulky appearance of edges.

· Reinforcing/backing upper components: more fragile areas or those that need certain stitching or holes require a bit of reinforcement, usually existing between the lining and the uppers.[6]

· **Seam rubbing**: to reduce the bulk of the opened **stitched seam**.

· **Preassembling**: prepositioning of upper components before sewing.

· **Embossing**: printing of specific patterns and fancies on the **grain side** of upper components.

· **Silk screen-printing**: for decoration purposes.

· **Upper crimping: pre-molding** of upper component particularly for boots.

Upper closing means to stitch the various cut sections to produce the complete uppers in readiness for lasting. This involves a considerable number of small operations varied and in different sequences for different types of uppers.

重点及难点句

1. Material exploitation/utilization is a matter of knowledge and it is the most important aspect of upper cutting for what concerns quality and material economy.

材料的开发 / 利用是一门学问，是帮面裁断工序中最重要的环节，因为这关系到质量和材料的节省。

2. For homogeneous materials the knowledge required is little because, when the best exploitation method has been determined for a specific product and a given material, all that has to be done is to follow the method.

对于均质材料，所需要的知识很少，因为当特定产品和给定材料确定了最佳排料方案时，所要做的就是遵循该方法。

3. No definite exploitation method can be determined, only general principles, which have to be applied to, as much as each piece of material permits.

没有固定不变的排料方案，只能确定在允许的情况下适用于每一块皮料的一般原则。

4. Upper making/closing is a succession of preparation and sewing operations of which sequential order will depend on the model of the upper or on the technical process.

制帮 / 缝合包含一系列的准备和缝纫操作，其先后顺序取决于帮面类型或工艺流程。

5. The top piece usually is thinned to half thickness while the bottom is "skived to zero".

通常上压件被片剖到一半的厚度，而被压件需要"片边出口"。

6. Reinforcing/backing upper components: more fragile areas or those that need certain stitching or holes require a bit of reinforcement, usually existing between the lining and the uppers.

帮面补强件：强度较低的区域或某些缝合或冲孔需要补强的部位，（补强材料）通常位于鞋里和鞋面之间。

词 汇

upper cutting：帮面裁断　　　　　　　cutting operation：裁断操作

material exploitation/utilization：材料的开发 / 利用；排料方案

irregularity：不规则性　　　　　　　cutter or clicker of leather：皮革裁断工

profit or loss：盈利或亏本

upper making/closing：制帮，帮部件的加工

splitting：片料，片剖，通片　　　　skiving：片边

stitch marking：点缝帮标志点

decoration seam：装饰缝　　　　　　decoration perforation：装饰花孔（眼）

identification marking/stamping：点标志点

size-fitting series No.：尺码型号　　thin：削薄

gimping：切锯齿形边口　　　　　　punching：冲孔

fitting fastening：装配紧固件　　　　hand stitching：手工缝

perforating：冲孔，打孔　　　　　　cementing：刷胶

adhesive：胶黏剂　　　　　　　　　edge folding：折边

reinforcement：补强材料，补强带　　edging：熨烫边口

raw edge：切割边，断口　　　　　　binding：沿口，滚边，包边材料

edge beating：锤敲边口　　　　　　seam rubbing：磨平针眼孔

stitched seam：合缝线　　　　　　　preassembling：镶接

embossing：压印　　　　　　　　　grain side：粒面

silk screen-printing：丝网印刷

upper crimping：（靴）帮面压翘、压凹

pre-molding：预成型　　　　　　　upper closing：合帮套，帮面缝合

Lesson 3 Cutting and Related Tools & Equipment

Inside the shoe factory, the first operations in the manufacturing process are the cutting and marking operations. The cutting operation can be done manually or mechanically.

1. Hand cutting

Cutting knives of different sorts and **cutting boards** made of hard plastic material can be used for upper cutting.

The knife blade has to be perfectly sharpened and particularly the blade tip should always be keen. The shaping and initial sharpening are done on both sides of the blade with a grinding wheel, and then a **sharpening stone** and a leather strap are used to maintain the blade's sharp curves.[1]

The material used for cutting board should be hard enough to produce a neat cut and soft enough to prevent early blunting of the blade.

2. Machine cutting

The **cutting presses**, also called **clicker cutting machines** are used and cutting dies for each shoe pattern part are required.

There are different sorts of cutting machines using cutting dies: **swing arm cutting press**, **beam cutting press**, **traveling beam cutting press**, **traveling head cutting press**, etc. Generally cutting presses are equipped with safety devices, such as double switch operation requiring both hands, or **photoelectric cell**.

Swing arm cutting press is mainly used to cut leather because the operator can see the surface and place the **cutting dies** in areas to obtain the greatest yield. In this operation, the operator works with a range of dies to cut out the various parts of a shoe. This permits the matching of parts for a pair from the same piece of leather.[2] **Clicker** operations are not fast. In fact, a line of several clickers is needed to meet production demands.

Large **table knife cutters** are used to cut multiple layers of fabrics. These machines are computer-controlled and offer very efficient cutting figures. These machines also have power-assisted movement of the materials so that after a lay has been cut, the material will advance so that the next lay can be cut.

A traveling head press is especially used to cut out multiple layers of **insole board** and other items on man-made materials where the volume of the run is usually long.[3]

Turret-head machines, which can carry six cutting dies are used to cut a wide range of materials that are mechanically advanced under the cutting head toward the operator who picks and sorts the parts.[4] Because this system can be fully automatic, from computer-generated programs to machine control, it offers those companies cutting man-made upper materials and board production numbers at a minimum cost.

Waterjet cutting has become an important tool for cutting leather. Under this system, the surface of the leather is scanned, and the parts are automatically **nested** by the computer to avoid **blemishes** and **defects** and placed in their correct cutting areas to obtain the best cutting yield.[5] In the waterjet program, the scanning and nesting program takes place on a table on one side of the **cutting chamber**. When the nesting program is completed, the table moves into the cutting chamber. When cutting is completed, the table moves to the far side of the cutting chamber. These systems are quite fast, and can cut relatively small parts. The problems of **splashback** noted in the earlier systems have been eliminated by the use of **glycerin** in the water to hold the water together.[6]

Laser cutting is fast. It travels at speeds up to three feet per second, depending upon the material being cut. Unfortunately, laser cutting leaves an unpleasant odor and some fumes. Laser cutting also leaves the edges of the materials with an **undesirable finish** and some **charring**.[7]

Vibrating knife digital cutting machine is a perfect combination of technique and technology. With this machine, rapid nesting and efficient cutting operations can be conducted while saving time, manpower and material. It works for a wide range of materials and is suitable for **sample cutting** and small bulk production.

The advantages include effectively identifying flaws in leather by using the projector or camera to capture leather contour; adjusting the cutting direction in accordance with the texture of natural leather material; carrying out interactive cutting, punching, and **drawing** in a single period of running time, etc.[8] All operations are computerized by using the computer to simulate the operation process exactly, which rules out the interference from the factors of the workers' mood, knowledge, tiredness and other personal factors that happened in traditional cutting, thus avoiding hidden waste and increasing the utilization rate of the materials.

重点及难点句

1. The shaping and initial sharpening are done on both sides of the blade with a grinding wheel, and then a sharpening stone and a leather strap are used to maintain the blade's sharp curves.

使用砂轮对刀片的两侧进行开刃和初步打磨，然后使用磨刀石和皮带来保持刀片的锋利曲线。

2. This permits the matching of parts for a pair from the same piece of leather.

这样可使来自同一块皮革的部件匹配成一双鞋。

3. A traveling head press is especially used to cut out multiple layers of insole board and other items on man-made materials where the volume of the run is usually long.

动头式裁断机特别适用于多层内底机的切割以及从人造材料上下裁其他部件，这些材料通常很长。

4. Turret-head machines, which can carry six cutting dies are used to cut a wide range of materials that are mechanically advanced under the cutting head toward the operator who picks and sorts the parts.

可以携带 6 个刀模的转塔头裁断机可用于各种材料的下裁，将切割头的下方的材料机械地朝着裁断工的方向输送，裁断工取出下裁好的部件并进行分类。

5. Under this system, the surface of the leather is scanned, and the parts are automatically nested by the computer to avoid blemishes and defects and placed in their correct cutting areas to obtain the best cutting yield.

该系统对皮革表面进行扫描，计算机对部件自动套排以避免瑕疵和缺陷，并将其放在正确切割区域以获得最佳的出裁率。

6. The problems of splash back noted on the earlier systems have been eliminated by the use of glycerin in the water to hold the water together.

早期（切割）系统中出现的回溅问题已经通过在水中加入甘油将水聚在一起而得以消除。

7. Laser cutting also leaves the edges of the materials with an undesirable finish and some charring.

激光切割还会使材料的边缘变得不整洁或碳化。

8. The advantages include effectively identifying flaws in leather by using the projector or camera to capture leather contour; adjusting the cutting direction in accordance with the texture of natural leather material; carrying out interactive cutting, punching, and drawing in a single period of running time, etc.

其优点包括使用投影仪或照相机捕捉皮革轮廓以有效识别皮革上的瑕疵；根据天然皮革材料的纹理调整切割方向；在同一运行时间内进行交互式切割、打孔、标记等。

词 汇

hand cutting：手工裁断　　　　　cutting knives：（手工）裁刀

cutting board：裁断垫板　　　　　sharpening stone：磨刀石

cutting press/clicker cutting machine：裁断机

swing arm cutting press：摇臂裁断机　　beam cutting press：横梁式裁断机

traveling beam cutting press：动臂式裁断机

traveling head cutting press：动头式裁断机

photoelectric cell：光电电池　　　cutting die：裁断刀模

clicker：裁断工，裁断机　　　　　table knife cutter：台式裁断机

insole board：内底板　　　　　　turret-head machine：转塔头裁断机

waterjet cutting：高压水束裁断　　nest：套排，嵌套

blemish：瑕疵　　　　　　　　　defect：缺陷

cutting chamber：切割台　　　　　splashback：（水流）溅射

glycerin：甘油   undesirable finish：较差的光洁度

charring：碳化

vibrating knife digital cutting machine：震动刀数字裁断机

sample cutting：样品切割   drawing：标记

CHAPTER 3

Lasting and Bottoming Process

Lesson 1　Lasting

Once the shoe upper is sewn, the next step in shoemaking is to pull the upper onto the last. In the shoe lasting, the main manual operations are the correct **positioning** of the both sole and the upper on the last, and feeding the lasting machine.

Lasting is the conversion of flat material into the form of a shoe upper. It involves pulling and stretching the upper. Dry heat or steam is often used to soften the uppers, getting them ready to stretch. The main aim is to get the flat upper material to conform as closely as possible to the last. Excessive straining of the upper material must be avoided, and the upper must be prevented from **reasserting** itself when the last is removed.

The upper of the shoe is tacked to the back of the last to ensure the back height is correct.[1] The upper must be pulled and stretched to take the shape of the shoe last. The first strain is in the lengthwise direction and it can be achieved by different means. Its purpose is to **bed** the **stiffener** to the last, tighten the **top line**, and position the toe of the upper accurately over the insole.[2]

This is often achieved nowadays by bedding the stiffener on a **back part forming machine** using a forward pull.[3] This should then allow the upper to be placed in a **forepart lasting machine** with the correct amount fed into the front pincers and with the topline already formed so that no excessive pull is executed on the forepart.[4] The latest range of pulling over and forepart lasting machines use computer controls where much of the skill in setting up the machine is taken away from the operator.[5]

The pull of the **pincers** is individually adjustable on most modern machines and the sequences, in which the pairs of pincers pull, can be controlled on most of the latest models, enabling shoes on lasts with **bump toes** to be successfully lasted. A Teflon band holds the upper firmly in place while the pincers release and **wiper plates** wipe the upper.[6] At the same time the **hot melt cement** applicator imprints hot melt adhesive on the insole, the upper and insole being bonded as the wiper plates move over, giving an excellent bond between the upper and insole.

The forepart lasting machine and the **waist** and seat lasting machine can be linked, so that one computer will set both machines. On some models, a programmed number can be called up which will set all the required adjustments for a particular shoe. Alternatively, some manufacturers use a barcode on the work ticket where the operator will wipe the code and the machine will be automatically set for that batch of shoes and materials being used. The linking together of these operations reduces the number of workers required and gives a much more consistent result.

After this operation is completed a simple robot can pick up the shoe and unload it to the next operation, which is **heat setting**. At last, the uppers are often cooled to shrink them down tight to the last before the shoe bottom can be attached.[7]

重点及难点句

1. The upper of the shoe is tacked to the back of the last to ensure the back height is correct.

鞋帮被固定在鞋楦后身，以确保后帮高度正确。

2. Its purpose is to bed the stiffener to the last, tighten the top line, and position the toe of the upper accurately over the insole.

其目的是使主跟贴伏在鞋楦后身、使鞋口绷紧并调整好鞋头（帮脚）在内底上的位置。

3. This is often achieved nowadays by bedding the stiffener on a back part forming machine using a forward pull.

如今，这通常是将主跟材料在后帮成型机上通过向前的拉伸来实现的。

4. This should then allow the upper to be placed in a forepart lasting machine

with the correct amount fed into the front pincers and with the topline already formed so that no excessive pull is executed on the forepart.

然后将鞋面放入绷前帮机中，前部的钳子夹持住合适的帮脚余量，同时由于鞋口已成型，这样就不会对前部施加过度的拉力。

5. The latest range of pulling over and forepart lasting machines use computer controls where much of the skill in setting up the machine is taken away from the operator.

最新的拉帮绷尖机系列使用计算机控制，大部分的机器设置都无需操作员来操控。

6. A Teflon band holds the upper firmly in place while the pincers release and wiper plates wipe the upper.

当松开夹钳、卡板推倒帮脚时，聚四氟乙烯卡带将鞋面牢牢固定。

7. At last, the uppers are often cooled to shrink them down tight to the last before the shoe bottom can be attached.

最后在合鞋底之前，鞋帮通常通过冷却收缩而紧附于楦体。

词 汇

position：定位   reassert：恢复原状

bed：使平铺于……上，使紧伏于   stiffener：主跟，定型材料

top line：鞋口，（靴子）筒口，楦头弧线

back part forming machine：后帮 / 主跟（预）成型机

forepart lasting machine：绷尖机   pincer：（绷帮机上的）夹钳

bump toe：高头（楦）   wiper plate：卡板

hot melt cement：热熔胶   waist：腰窝部位

heat setting：加热定型

Lesson 2 Bottoming

When the upper was lasted the next operation happens fast. The workers pull a shoe off the line and match it to the correct size outsoles. A **foot-activated press** holds the shoe down while the worker marks the top edge of the sole with a pen. This mark is called the glue line. In this case, the line is a guide for the workers at the cementing workstation. If this shoe required **roughing**, the workers with the buffing tools would follow the same line before the **primer** and glue were applied.[1]

Next, the workers apply the primer and cement to the uppers and the shoe outsoles. The outsoles and uppers are on opposite sides of the **assembly line**. This allows the workers to apply different compounds of primer and glue to each part. The outsole requires a different primer for a strong bond.

At the next station, the outsoles and uppers are moved to the conveyor belt. This allows the glue to be dried at the different temperature.

Once the primer and cement are dry, the parts can be assembled. The worker matches the correct size outsole with the upper and then presses them together by hand. They align them carefully but quickly, starting at the toe to make sure it's centered. Next, the worker will flip the shoe and set the heel into the position. The cement is dry but **tacky** so it can be repositioned without affecting the final bonding strength.[2]

A metal or plastic tool is used to make sure the sole follows the glue line and confirm the edges are straight.[3] This tool can also be used to remove air bubbles or unroll any tucked edges on the sole.[4]

To speed set the glue bond, the shoes are cooled in a chilling tunnel. To ensure the outsole and upper have complete contact, the shoe will go through the pressing operation. The bottom press should be made with a matching negative casting of the outsole or a contoured pressing plate to ensure the arch area is fully pressed.[5] After pressing, it's time to remove the shoe last and insert the footbed.

Another solution for attaching soles is to inject them directly into the upper part using molds with unique cavities.[6] Injection molds are manufactured for each sole design, for each foot, right and left, and for each size number.

The shoe upper is placed onto the upper movable mold part and then the mold slowly descends to the lower embossing tool.[7] After setting the boot upper onto the lower mold part, the tool closes completely and the polyurethane system is injected to form a sole. Depending on the applied PUR system, the size of the sole and the equipment used, the injection time is from 15s to 40s. The temperature of the tool and the **curing time** also depend on the system used and the size of the polyurethane sole. The **tool temperature** is usually from 25°C to 95°C and the time of cementing the PUR sole with the shoe upper in a compact unit is from 5mins to 8mins. After this time, the tool is opened and the boot is removed from the tool in about 10s; the tool is then ready for the next cycle of this process.

重点及难点句

1. If this shoe required roughing, the workers with the buffing tools would follow the same line before the primer and glue were applied.

如果鞋子（的帮脚）需要砂磨，在刷处理剂和胶黏剂之前，抛光的工作人员要先按照这条线迹进行打磨。

2. The cement is dry but tacky so it can be repositioned without affecting the final bonding strength.

胶是干的但有黏性，因此可以在不影响最终黏结强度的情况下重新定位。

3. A metal or plastic tool is used to make sure the sole follows the glue line and confirm the edges are straight.

使用金属或塑料工具来确保鞋底沿着胶线，并确认外沿是直的。

4. This tool can also be used to remove air bubbles or unroll any tucked edges on the sole.

这个工具也可以用来去除气泡或展开鞋底上的褶皱边缘。

5. The bottom press should be made with a matching negative casting of the outsole or a contoured pressing plate to ensure the arch area is fully pressed.

外底压合机装配了与大底配套的负模或廓形压板，以确保足弓区域得到充分的压合。

6. Another solution for attaching soles is to inject them directly into the upper

part using molds with unique cavities.

固定鞋底的另一种解决方案是将鞋底材料直接注入具有独特空腔的模具（与鞋面结合部分）中。

7. The shoe upper is placed onto the upper movable mold part and then the mold slowly descends to the lower embossing tool.

鞋帮放置在可移动的上模中，然后缓缓下降到带底纹的下模处。

词　汇

foot-activated press：脚踏式压机（画线机）

roughing：砂磨

primer：处理剂

assembly line：装配线

tacky：发黏的，黏性的

curing time：硫化时间，（交联）反应时间

tool temperature：模（具）温（度）

Lesson 3   Outsoles

The outsoles for running shoes, biker boots, and **football cleats** all have very different performance requirements and manufacturing methods. Performance factors for footwear outsoles include **traction**, support, **flexibility**, weight, slip resistance, and **durability**.

1. Performance factors for outsoles

Outsole traction is an important feature of any shoe outsole design. The design of the outsole pattern and selection of materials control the amount of traction. The traction requirements for hiking boots, office shoes, boat shoes, and bowling shoes are all radically different. When designing an outsole, it is important to understand the environment and surfaces the outsole will encounter. Traction is also directional. A mountaineering boot will require lateral traction on rough terrain, while a basketball sole needs to support quick stops on a smooth wood surface and allows spins with the foot planted.[1]

An outsole design must have some supportive features to ensure users can walk or run safely. A fast, lightweight running shoe designed for speed may have little support, while a mountaineering boot will have a **metal shank** to carry the weight of a heavy pack in rough terrain.

Depending on the intended purpose of the shoe, the outsole will require more or less flexibility. A tennis shoe or running shoe must be flexible, while a cowboy boot will have a steel shank to spread the load of the stirrups without bending at all. While more flexibility is usually a good thing, too much flexibility can cause instability and lead to foot fatigue. The general rule is that the heavier the load and the rougher the terrain, the stiffer the outsole.

Depending on the intended purpose of the shoe, the weight of the outsole may be a key feature. For a long-distance racing shoe or **track spike**, every gram is critical. For a driving shoe or biker boot, a heavy outsole is not a problem.

The durability requirements of an outsole design will depend on the environment and tasks selected for the particular shoe. The fine leather soles of women's dress shoes and men's office shoes are perfect for smooth stone hallways and carpeted offices but would last only a few steps on a rainy construction site or mountain trail. Durability can be a selling feature for a shoe but may come at the expense of added weight or reduced flexibility.

Slip resistance is a key feature of service shoe outsole design. Restaurant, hospital, maintenance, and warehouse staff members are required to wear shoes with **certified** slip-resistant rubber compounds and **tread patterns**. Oil-resistant rubber compounds may also be a requirement for industrial footwear.

2. How to make shoe sole

The outsole is one of the most expensive parts of the shoe. For low-cost shoes, the cost of an outsole can be 10% of the total cost, second only to the leather cost. For high-end basketball, running, or soccer shoes the outsole can be 25% to 35% of the shoe's factory price.

Rubber cupsole: This style is called a "cup" sole because it "cups" the upper of the shoe. The cupsole is a very common outsole and can be found in many styles of footwear. Hiking boots, casual shoes, army boots, skate shoes, court shoes, and sometimes a stylized cupsole can be found on inexpensive running shoes. A cupsole may have a "drop-in" midsole made of EVA blocks or an "egg crate" midsole molded inside the outsole. A cupsole is made by compression molding **uncured** rubber into a mold, much like making waffles.[2]

PU outsole: Soles of polyurethane systems are produced by liquid injection molding (LIM) technology. This process includes weighing, mixing and pouring two liquid plastic components. This technology is different from the standard process of reaction injection molding because it is based more on mechanical agitation than on forced stirring under pressure.[3] In this way, a large number of parts with different characteristics can be produced.

Compression molded EVA outsole: This running shoe outsole design uses a dual-density compression molded EVA midsole with rubber inserts and a carbon fiber shank.[4] This is an expensive outsole to make as it requires molds for the EVA, rubber, and shank parts. The EVA midsole is made by first bonding two EVA preforms with the molded shank in between.[5] The final shape is made by compression molding. The compression molded rubber parts are glued on in a separate operation.

Classic vulcanized outsole: In **vulcanized** shoe making process, the grey rubber bottom is cemented to the upper, and then the rubber foxing tape wraps both.[6] A toe cap and **toe bumper** are also added. Once the assembly operation is completed, the entire shoe is **cooked** at 110°C for 70 minutes to **cure** the rubber; this makes the bonds permanent. This outsole requires a rubber compression mold for the bottom and a vulcanizing production factory to make the rubber parts.[7]

Injection molded outsole: **Cleated shoes** for soccer, football, and baseball require stiff supportive bottoms. This outsole requires a plastic injection mold. Complicated designs with several over-molded colors require an expensive mold. Simple designs require a less expensive mold. Cleats with metal spikes or inserts require over molding or insert molding processes.[8] This type of plastic injection requires a specialized molding machine and is done outside of the shoe factory by a subcontractor.

重点及难点句

1. ... a basketball sole needs to support quick stops on a smooth wood surface and allows spins with the foot planted.

……篮球鞋底需要能在光滑的木质地面上快速停止，并允许在脚着地时旋转。

2. A cupsole may have a "drop-in" midsole made of EVA blocks or an "egg crate" midsole molded inside the outsole. A cupsole is made by compression molding uncured rubber into a mold, much like making waffles.

成型底可能有一个由 EVA 块制成的"嵌入式"中底或一个模制在外底内的"鸡蛋饼"一样的中底。成型底是通过将未硫化的橡胶在模具中模压成型的，很像制作华夫饼。

3. ... it is based more on mechanical agitation than on forced stirring under pressure.

……它更多基于机械搅拌而不是在压力下的强制搅拌。

4. This running shoe outsole design uses a dual-density compression molded

EVA midsole with rubber inserts and a carbon fiber shank.

这种跑鞋外底设计与双密度模压成型 EVA 中底、橡胶衬垫和碳纤维板勾心共同使用。

5. The EVA midsole is made by first bonding two EVA preforms with the molded shank in between.

EVA 中底是通过先将两个 EVA 预成形件黏合在一起而制成的，中间夹着模制勾心。

6. In vulcanized shoe making process, the grey rubber bottom is cemented to the upper, and then the rubber foxing tape wraps both.

在硫化制鞋过程中，将未硫化的橡胶底与鞋帮黏合，然后用橡胶胶条将两者包裹起来。

7. This outsole requires a rubber compression mold for the bottom and a vulcanizing production factory to make the rubber parts.

（制备）这种外底需要使用橡胶压模，硫化厂制造橡胶部件。

8. Cleats with metal spikes or inserts require over molding or insert molding processes.

带有金属钉或内插的钉鞋底需要包覆成型或内插成型工艺。

词 汇

football cleat：足球（钉）鞋

flexibility：弹性，柔性

metal shank：金属勾心

certified：达到标准的，有合格证书的

rubber cupsole：橡胶成型底

vulcanized：硫化的

cure：硫化

injection molded：注塑的（外底）

traction：抓地力；牵引力

durability：耐久性

track spike：田径鞋，跑鞋

tread patterns：鞋底花纹

uncured：未硫化的，未固化的

toe bumper：（硫化鞋）前围条

cook：热硫化

cleated shoe：钉鞋，防滑鞋

CHAPTER 4

Shoe Construction

Lesson 1　Methods of Shoe Construction

Shoe construction refers to the method used to attach the upper to the **lower**. After the various components of the shoe's upper have been cut, sewn, and stretched over the last, the components of the lower can be attached using one of several methods. There are many ways to attach the sole to the upper but commercially only a few methods are preferred. The process of **bottoming** will determine the price, quality and performance of the shoe.

1. Board lasted construction

Alternative names include **stuck on construction**, **cement-lasted construction and the Compo Process**. It is a very common shoe construction technique that can be made by machine or by hand. It involves pulling the upper over the last and cementing the **underflaps** to the insole board.[1] The outsole and midsole or heel are then cemented to the bottom of the insole board. Because the insole board can be observed when looking inside the shoe, the result is often termed a board-lasted shoe. Board lasting construction can be used in almost any style of shoe, being suitable for many upper material types and outsole styles, especially for lightweight and flexible footwear.

2. Goodyear welt construction

Goodyear welt refers to a technique developed in the late nineteenth century with the advent of the Goodyear sole-stitching machine.

The "welt" is a strip of leather which is sewn around the bottom edge of a shoe. This stitching (the welt seam) attaches the welt to both the insole and the upper of the shoe.

The welt is folded out to form a point of attachment for the outer sole.[2] The outer sole (two layers can be used in heavier shoes) is sewn to the welt, with a heavy-duty **lock-stitch** seam.

Crucially, this stitching runs around the outside of the sole (rather than piercing the part under the foot) to maximize the sole's water resistance. In contrast, the **Blake construction** (a widely used method for making formal shoes) involves stitching the outer sole directly to the insole, resulting in a seam that can sometimes be felt inside the shoe, and which is more likely to **leak**.

In Goodyear welt construction, there is a thin cavity between the insole and the outer sole. We fill this with cork, a material that is lightweight, insulating, molds to the shape of the foot and, most importantly, breathes.

The disadvantage of the welted construction is that it can add bulk to the shoe, as its outer edges need to be wide enough to accommodate stitches. However, there are several significant advantages, in particular: they can be repaired more easily and they are more weather-resistant.

Bond welt is a variation with its distinguishing feature being a strip of welting attached by stitching or cementing to the top edge of the insole.[3] The shoe is then flat lasted. This is not a true welt construction.

3. Stitchdown construction

A cheaper method used to produce lightweight flexible soles for children's shoes and some casual footwear describes the upper turned out (flanged) at the edge of the last. This is then stitched to the **runner**. In some countries it is known as "veldt" and "veldtschoen".

4. Moccasin construction

Thought to be the oldest shoe construction this consists of a single layer section which forms the insole, vamp and quarters.[4] The piece is molded upwards from the under surface of the last. An **apron** is then stitched to the gathered edges of the vamp and the sole is stitched to the base of the shoe. This method is used for flexible fashion footwear.

The **imitation moccasin** has a visual appearance of a moccasin but does not have the **wrap around construction** of the genuine moccasin.

5. Molded construction

The lasted upper is placed in a mold and the sole is formed around it by injecting liquid synthetic soling material (PVC, polyurethane).[5]

Alternatively, the sole may be **vulcanized** by converting **uncured rubber** into a stable compound by heat and pressure. When the materials in the molds cool the sole-upper bonding is completed. These methods combine the upper permanently into the sole and such shoes cannot therefore be repaired easily. Molded methods can be used to make most types of footwear.

6. Force lasting construction

Alternative names include: The **Strobel-stitched method** (or **sew-in sock**)

Force lasting has evolved from sports shoes but it is increasingly used in other footwear. A force-lasted shoe has no insole board. The upper is sewn directly to a sock by means of an **overlocking machine** (Strobel stitcher).[6] The upper is then pulled (force lasted) onto a last or molding foot. **Unit soles with raised walls** or molded soles are attached to completely cover the seam. Some models of athletic footwear are half slip-lasted, which in most cases means that the shoe is board-lasted (or cement-lasted) in the heel and **shank sections** and slip-lasted in the forefoot.

This technique is sometimes known as the Californian process or slip lasting.

重点及难点句

1. It involves pulling the upper over the last and cementing the underflaps to the insole board.

（绷楦法）是指通过拉伸将鞋帮绷附于楦体并将帮脚黏合在内底板上。

2. This stitching (the welt seam) attaches the welt to both the insole and the upper of the shoe. The welt is folded out to form a point of attachment for the outer sole.

此项缝合（缝沿条）将沿条与内底及鞋帮同时缝合在一起。沿条的向外延伸提供了与外底的结合点。

3. Bond welt is a variation with its distinguishing feature being a strip of welting attached by stitching or cementing to the top edge of the insole.

假沿条鞋是（缝沿条鞋的）一种变体，其显著特征是使用缝合或黏合的方式将沿条固定在内底边缘上。

4. Thought to be the oldest shoe construction this consists of a single layer section which forms the insole, vamp and quarters.

这被认为是最古老的鞋子结构，它由单层材料构成内底、前帮及后帮。

5. The lasted upper is placed in a mold and the sole is formed around it by injecting liquid synthetic soling material (PVC, polyurethane).

绷帮成型后的鞋帮放置在模具中，并通过注入液体合成材料在其周围形成鞋底。

6. The upper is sewn directly to a sock by means of an overlocking machine (Strobel stitcher).

鞋面是通过锁缝缝纫机直接缝到垫式内底上的。

词 汇

lower：底部件　　　　　　　bottom：合底，制底

board lasted (stuck on，cement-lasted) construction/ Compo Process：绷楦结构，绷楦法

underflap：帮脚，绷帮余量　　Goodyear welt construction：固特异缝沿条结构

lock-stitch：锁缝　　　　　　Blake construction：缝内线结构

leak：渗入，透水　　　　　　bond welt: 假沿条（鞋）

stitchdown construction：帮脚外翻线缝结构，拎面结构

runner：鞋底边沿　　　　　　aporn：前帮盖

imitation moccasin：仿莫卡辛鞋

wrap around construction：（帮料）包底结构

molded construction：模压工艺

vulcanized：硫化的

uncured rubber：未硫化橡胶

force lasting construction (Strobel-stitched method，sew-in sock)：闯楦工艺

overlocking machine：锁缝机

unit soles with raised walls：带边墙的成型底

shank section：腰窝部位

Lesson 2   Vulcanized Shoe Making Process

Vulcanization is the process of heating raw rubber to **cure** it. This process creates **crosslinks** inside the rubber compound bonding it together. Before the rubber is vulcanized it is **stretchable**, **gummy**, and easy to tear. After being vulcanized the rubber is tough, stretchable, and ready to wear.[1]

Vulcanized shoe construction is a much older technology. In the 1980s, modern cold cement construction replaced vulcanized construction. However, fashion changes and the vulcanized shoe made a huge comeback in 2010. Due to the high temperature required to vulcanize, or cook the rubber outsole, uppers must be made from heat-resistant materials, commonly canvas or suede leather.[2]

1. Vulcanized shoe construction

If a shoe is to be vulcanized then we need to use a metal last. Plastic last cannot survive the heating required to cure the rubber.

Before assembly can start, all of the outsole components must be prepared. The uncured rubber **foxing tape** must be made just before assembly. If the rubber parts **age**, they will not bond correctly to the other shoe parts. To apply the foxing tape, we will need the actual bottom of the shoe. This part has already been molded, it's about 90% vulcanized so it's still a little bit soft and can be bonded in the assembly process.

This part will also have the **midsole filler** added. The filler material is gray rubber with some air bubbles blown inside, and it's made from recycled uncured foxing tape. With vulcanized construction, we cannot use EVA foam because the heat of the vulcanizing process will destroy the foam.

2. Board lasting operations

In vulcanized construction, the last must be aluminum to survive the oven temperature. Metal lasts also heat up and cool down quickly. The first step is to lightly cement the lasting board to the bottom of the last, just enough to hold it in place during the lasting operation. Lasting boards are made from different materials and can be stiff or flexible, thin, or thick.

Now, the last and upper are taken to the lasting machine. The toe lasting machine pulls the upper down onto the last and securely bonds the two parts together in one operation. Once the shoe heel and waist of the shoe are lasted, the upper is ready for the outsole.

The first step of the vulcanized outsole assembly process is similar to the cold cement process. The upper and the rubber sole parts all receive their coating of primer and cement. Now, the rubber bottom with the **cushioning wedge** is bonded together with the upper.[3]

The foxing tape covers the rubber outsole part and overlaps up onto the upper. The foxing tape must cover 5mm of the upper to have a solid bond. The shoe can now have the extra **toe tape** added, and then a rear logo will be applied to cover the joining seam.

With the tape applied, it is time to rip off any extra rubber using a hot knife and to make sure there are not any gaps.[4] The sole will get a quick pressing to make sure the parts are fitting correctly.

3. The vulcanizing oven

With the last still inside, the shoe is placed on a steel rack so it can be wheeled onto the oven. The shoe will be "cooked" in the vulcanizing oven for several hours. The shoe is heated long enough for the uncured foxing tape and the sole unit to fuse together.

After cooling, the shoe last is removed, the **footbed** is inserted, and the laces are attached. The shoes are now ready for final inspection, cleaning, and packing.

重点及难点句

1. Before the rubber is vulcanized it is stretchable, gummy, and easy to tear. After being vulcanized the rubber is tough, stretchable, and ready to wear.

硫化前橡胶是可拉伸的、有黏性的，并且容易撕裂。硫化后的橡胶坚韧、可拉伸，而且耐磨。

2. Due to the high temperature required to vulcanize, or cook the rubber outsole, uppers must be made from heat-resistant materials, commonly canvas or suede leather.

由于橡胶外底硫化需要高温，鞋面必须由耐热材料制成，通常是帆布或反绒面革。

3. The upper and the rubber sole parts all receive their coating of primer and cement. Now, the rubber bottom with the cushioning wedge is bonded together with the upper.

鞋面和橡胶鞋底都刷了处理剂和胶黏剂。现在，带有缓冲内插的橡胶底与鞋面黏合在了一起。

4.With the tape applied, it is time to rip off any extra rubber using a hot knife and to make sure there are not any gaps.

贴好外胶条后，使用热刀切除多余橡胶胶料并确保没有缝隙。

词 汇

vulcanization：硫化
crosslink：交联
gummy：有黏性的
age：老化
cushioning wedge：缓冲内插（楔状物）
vulcanizing oven：硫化罐

cure：硫化，熟化
stretchable：可拉伸的
foxing tape：外胶条，围条
midsole filler：中底填充物
toe tape：胶包头
footbed：鞋垫

Lesson 3   The Sewn Shoes

Depending on the function and style of the shoe design, there are many options of connecting the upper and the outsole unit.

Whether you are designing running shoes, Alpine mountaineering boots, or posh office shoes, you will need an understanding of shoe construction methods. Here we will review these common forms of sewn shoes including Blake construction, Blake/rapid construction, Goodyear welt construction, Norwegian storm construction, Stitch down construction, and **Bologna constructions**.

Footwear construction terms

Welt: A welt is a strip of leather, rubber, or plastic that runs along the perimeter of a shoe outsole. The machinery used for this process was invented in 1869 by Charles Goodyear Jr., the son of the famous rubber inventor, Charles Goodyear Sr.

Outsole: Can be made of rubber or leather.

Midsole: Can be made of leather, cork, or foam.

Common shoe construction styles

1. Blake construction (also called McKay welt)

Used to make flexible leather shoes, the Blake construction starts with a board-lasted upper. Glue the sole in place then with the last removed, sew the upper directly through the outsole unit.[1] The outsole may be leather or rubber and have a **groove** molded into its surface to guide the **Blake stitch**. You will find Blake construction on **weltless** leather dress shoes, moccasins, and boat shoes. Blake construction is not waterproof.

2. Blake/rapid construction

It is similar to the standard Blake construction, but with a "rapid" perimeter stitch attaching the outsole. The outsole covers the Blake stitching. This allows for a thicker sole and is easier to **resole**. The extra layer can be rubber, making the shoe more durable.

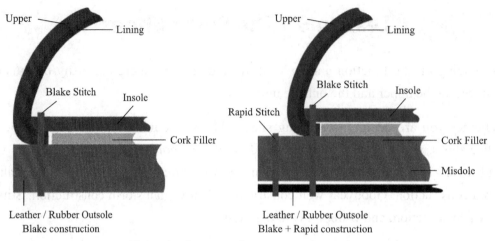

Figure 1　Common shoe construction styles

3. Goodyear welt construction

The Goodyear welt is often used to make waterproof soles, the stitch that attaches the sole to the shoe runs around the outside edge and does not make stitch holes in the upper.[2] The upper is sewn to the welts that attach to the insole and the outsole. During assembly, the welts are attached first by a horizontal "Goodyear" stitch, named for the inventor of the stitching machine, Charles Goodyear Jr. The Goodyear welt construction method is ideal for heavy-duty boots for hiking or service.

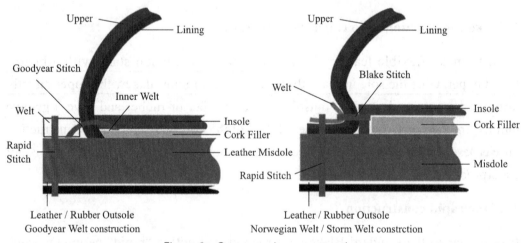

Figure 2　Common shoe construction

4. Norwegian welt / storm construction

Similar to the Goodyear welt construction method the upper is turned outside and is sandwiched between the outer welt and the outsole.[3] Used to make the heaviest waterproof boots, the Norwegian Storm Welt is difficult to make and is found almost exclusively in the workshops of Italian bootmakers.

5. Stitch down veldtschoen welt construction

The Veldtschoen welt is related to the Goodyear and Norwegian welt constructions. In this case, the rapid stitch line sewn through the midsole is paired with a second rapid stitch that attaches the outsole. The outsole is attached after the first welt, the bottom stitch is protected by the outsole.[4]

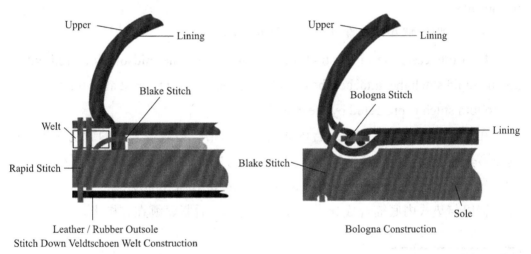

Figure 3   Common shoe construction styles

6. Bologna construction

Developed in Italy and used primarily for dress or fashion shoes, the Bologna construction creates a smooth, comfortable shoe. The shoe lining is joined into a sock fitting the last tightly.[5] The leather upper is then attached to the sole via the Blake stitching method. The Bologna shoe construction method is ideal for making very clean-looking and flexible shoes.

重点及难点句

1. ... the Blake construction starts with a board-lasted upper. Glue the sole in place then with the last removed, sew the upper directly through the outsole unit.

……缝内线结构首先进行绷楦，黏合外底后脱楦，然后将鞋帮（帮脚）与外底（透缝）缝合。

2. The Goodyear welt is often used to make waterproof soles, the stitch that attaches the sole to the shoe runs around the outside edge and does not make stitch holes in the upper.

固特异沿条结构通常用于制作防水鞋底，连接鞋底与鞋面的缝线环绕外侧边缘，不会在鞋面上留下缝线孔眼。

3. ... the upper is turned outside and is sandwiched between the outer welt and the outsole.

……帮脚外翻并被夹在外沿条和外底之间后缝合。

4. In this case, the rapid stitch line sewn through the midsole is paired with a second rapid stitch that attaches the outsole. The outsole is attached after the first welt, the bottom stitch is protected by the outsole.

在这种情况下，第一道缝合内底的缝线与第二道缝合外底的缝线平行，在缝完第一道缝线后再缝合外底，这样缝合内底的缝线就得到了外底的保护。

5. The shoe lining is joined into a sock fitting the last tightly.

鞋里与垫式内底缝合成袜子一样的"鞋套"后紧紧绷在鞋楦上。

词 汇

Bologna construction：博洛尼亚结构

Blake (McKay welt) construction：缝内线结构

groove：容线槽

Blake stitch：（缝）内线

weltless：无沿条的

Blake/rapid construction：内外线透缝结构

resole：更换外底

Goodyear welt construction：固特异缝沿条结构

Norwegian welt / storm construction：压条结构

stitch down veldtschoen welt construction: 双线透缝沿条结构

PART IV
Materials and Quality Inspection for Shoes

CHAPTER 1

Materials Used for the Upper Parts

Lesson 1   The Most Common Materials Used to Make Shoes

A huge part of learning how to design shoes is understanding how and why to select particular materials for the shoes. The material choice will decide how expensive or inexpensive these shoes will be in the store. The material choice will also affect **flexibility**, **durability**, comfort and import duties. Designers need to understand the limitations of leather, textiles, **synthetics**, **foam** and rubber as they relate to shoe design. The materials selected will have an impact on the **fabrication techniques**, stitching procedures, **reinforcements**, and even the type of glue used to assemble the outsole to the upper.

The most crucial feature in shoe design is not the pattern used for the shape and the look of the shoe but the fundamental material from which the shoe is made. The five materials most commonly used in shoe production are leather, textiles, synthetics, rubber and foam.

1. Leather

Leather is flexible yet durable, as sturdy as it is supple. It's elastic, so it can be stretched yet it resists tearing and abrasion. It's a **breathable** material, and it **insulates** heat, helping to regulate temperature. This all makes leather shoes conform to the feet as no other shoe material can.[1] It's no surprise that leather is one of the most common materials that shoemakers use, particularly in making dress shoes.

But leather does have some drawbacks. It can be heavy, hot, and susceptible to water absorption and damage if not treated well. **Water-resistant** and **water-proof** treatments add cost. Leather is a relatively expensive material when compared to fabric or other man-made materials and must be treated with care during shoe manufacturing. Because leather hides are from individual animals, each is a different size and each will have scars, **imperfections**, and even brands that must be avoided when cutting. This uncut material is called **cutting loss**. For leather, cutting loss is at best 5% of a hide, for the highest quality shoes, shoe leather cutting loss can be 15%. That's 15% of the material cost is thrown away.

2. Textiles

Footwear textiles come from many fiber types including cotton, wool, nylon, **polyester**, **polypropylene**, **rayon**, **Lycra** and many others. Each has its own look and physical properties. The most common textiles used for making shoes are cotton, polyester, wool and nylon. Nylon and polyester are lightweight and durable. Lycra is stretchable and cotton canvas is a must for vulcanized construction and has a look all its own.

One key advantage of textile shoes is their versatility in styles and designs. Each textile also has its own physical properties that must considered when choosing whether or not to own or wear it, such as variations in breathability, support and temperature control (i.e. hotness or coolness).[2]

3. Synthetics

Synthetic materials go by many different names—PU leather or simply PU, synthetic leather or simply synthetics. These man-made materials are often a composite made of two layers, being a backing layer made of woven or non-woven polyester fibers, combined with an external surface by a "dry" lamination process or by liquid "wet" processes.[3] Like their textile counterparts, synthetic materials also come in a variety of colors and textures.

The advantage of shoes made from synthetics is that they tend to be comparatively quite inexpensive for both the manufacturer and consumer as compared with other types of shoes. Because the material is less durable than its leather and textile counterparts, however synthetic shoes tend to degrade faster and need replacing more often.

4. Rubber

The outsole or sole unit is the bottom component of a shoe that provides **grip and traction**.[4] The outsole is commonly rubber but can be high-density PU or EVA foam. Dress shoes may have leather bottoms. While most shoe rubber over the years has been made from polyester, a more **environmentally sound** natural rubber is increasing in popularity as an alternative.

5. Foam

Foam is the most common material used to provide support in the uppers of shoes of all sorts. They're all generally divided into two categories: open and closed cells. In open cell foam, the material is **permeable**, allowing water and air to pass through; in closed cell foam, these open cells are sealed, preventing the gasses inside them from escaping.

Open cell foam is made of **polyurethane** plastic and tends to be softer than closed cell foam. Open cell foam can often be found in the **collars** and tongues of shoes, and thin sheets of PU foam are used to back fabric in most shoe uppers.

Closed-cell foam is generally denser.[5] Midsoles of shoes are all made from closed cell foam. Common closed cell foams include EVA (ethyl vinyl acetate), PE (Polyethylene), **SBR (Styrene butadiene rubber)**, PU (Polyurethane), Latex, and Neoprene, each with their own properties. EVA foam is used for backing mesh materials, Neoprene and SBR are used when elastic properties are required, while Latex is common for collar linings.[6] PE foam is very light but not so durable, making its use more limited.

重点及难点句

1. This all makes leather shoes conform to the feet as no other shoe material can. 这些性能使得皮鞋能够提供其他材质鞋所无法提供的舒适感。

2. Each textile also has its own physical properties that must considered when choosing whether or not to own or wear it, such as variations in breathability, support and temperature control (i.e. hotness or coolness).

每种纺织品也有其自身的物理特性，在选择是否选用时必须考虑到这些特性，例如透气性、支撑性和温度控制（即热或冷）的变化。

3. These man-made materials are often a composite made of two layers, being a backing layer made of woven or non-woven polyester fibers, combined with an external surface by a "dry" lamination process or by liquid "wet" processes.

这些人造材料通常由两层材料复合而成，由纺织或无纺布聚酯纤维制成的背衬层通过"干法"层压工艺或液体"湿法"工艺与外部表面结合。

4. The outsole or sole unit is the bottom component of a shoe that provides grip and traction.

外底或组合底是鞋的底部组件，提供抓地力和牵引力。

5. Closed-cell foam is generally denser.

闭孔发泡材料通常密度较大。

6. EVA foam is used for backing mesh materials, Neoprene and SBR are used when elastic properties are required, while Latex is common for collar linings.

EVA 发泡材料用于背层衬网布材料，当对弹性特性有一定要求时可以使用氯丁橡胶和丁苯橡胶，而乳胶是常见的鞋口衬材料。

词 汇

flexibility：屈挠性，柔韧性，柔软度　durability：耐用性，耐用度
synthetics：合成材料　　　　　　　　foam：海绵，发泡材料
fabrication technique：制造技术，制作工艺
reinforcement：定型材料，补强衬料，增（补）强（加固）件
breathable：透气的　　　　　　　　　insulate：绝缘，隔离，隔热
water-resistant：拒水的　　　　　　　water-proof：防水的
imperfection：缺陷　　　　　　　　　cutting loss：裁断损耗
polyester：聚酯，涤纶　　　　　　　　polypropylene：聚丙烯，丙纶
rayon：人造丝，人造纤维　　　　　　　Lycra：莱卡
grip：抓地力　　　　　　　　　　　　traction：牵引，牵引力
environmentally sound：环境友好的，环保的

permeable：可渗透的，可透过的　　polyurethane：聚氨酯

collar：护口，鞋口，装饰性沿口

SBR（Styrene butadiene rubber）：丁苯橡胶

Lesson 2   Textiles for Shoe Design

When considering any textile for your shoe design there are five things to consider: the **thread size**, fiber composition, weave pattern, backing material, **sizing**, and surface treatments.[1]

Thread size: The basic building block for fabric is, of course, thread. **Denier** is how thread weight is measured. 1 denier = 1 gram per 9000 meters of thread. Typical deniers are 110D for very lightweight fabric, 420D to 600D are common in shoes, and 1, 000D for boots and bags.

Fiber types: Footwear textiles come from many fiber types including cotton, wool, nylon, polyester, polypropylene, rayon, and Lycra. Each has its own look and physical properties like water absorption, **stretchability**, UV resistance, and **colorfastness**.

For shoe design, polyester and nylon are very common. Stretchable Lycra is often used for bindings and linings. Cotton is a must for vulcanized shoes as synthetic fibers tend to melt. Natural fibers like cotton or wool will accept **finishing** treatments. Cotton canvas shoe uppers can be salt or stone washed before assembly to give the shoes a special character.[2] Cotton can also accept an oiled or waxed finish, but this must be done after the shoe is assembled. Oily or waxed canvas cannot be easily bonded to the shoe outsole during assembly.

Fabric weaves: There are many ways to "weave" the fibers together. In a woven pattern, two fibers cross each other. The fibers running the length of the fabric are called the "**warp**". The fibers running across the fabric side to side are called the "**weft**". There are many weaves: **plain**, **twill**, **satin**, **basket**, **doddy**, and **ripstop**.

The "knit" is the other common way fibers are joined. In knitted fabrics, the thread follows a meandering path forming **symmetric** linked loops. These linked and meandering loops can be easily stretched in different directions giving knit fabrics more elasticity than woven fabrics.[3] Depending on the fiber type and knitting pattern, a knit fabric can stretch as much as 500%. Common knit types are **jersey**, **interlock**, **double knit**, and ribbed.[4]

High-tech "air" mesh or 3D mesh is made by knitting. Also known as sandwich mesh, the inner surface can be smooth and act as the shoe lining.

Fabric backing and sizing: Once the fibers are knit or woven, the fabric must be dyed, sized, and backed before it can be used in shoes. The freshly made fabric is soft and shapeless, not suitable for the use in shoes. It's the sizing and backing treatments that give the fabric the toughness and body to make it useful.[5] Sizing is a liquid resin treatment applied to the fabric. The fabric is stretched, heated, and treated with the sizing resin which holds the fibers in place.

The backing material is critical to the character of the fabric. There are two common backing types. The thinner clear **coating** is called PU; this is the cheaper, lighter, less waterproof coating. You can see the fiber under the coating. Next, we have PVC backing which is more solid, and you cannot see the fibers through the backing. PVC is used to make a very sturdy waterproof fabric.

Fabric surface treatments: There are many treatments for fabric. DWR (Durable Water Resistant) coating is common. Another treatment is called **brushing**. Nylex and Visa, or Visa Terry, are knit products where one side is brushed to tease up the soft fibers.[6] These are the most common shoe lining materials for sports shoes.

Fabric lamination: When the fabric is assembled into shoes it is often laminated with a thin layer of PU foam. The foam backing controls **wrinkles** and makes the fabric easier to handle during assembly. The foam also prevents inner layers from X-raying through the thin fabric. The fabric has **tricot** material laminated to the back of the foam.

重点及难点句

1. When considering any textile for your shoe design there are five things to consider: the thread size, fiber composition, weave pattern, backing material, sizing, and surface treatments.

在设计鞋子时若你打算使用纺织品材料，有五件事要考虑：纱线尺寸、纤维成分、编织图案、背衬材料、施胶和表面处理。

2. Cotton canvas shoe uppers can be salt or stone washed before assembly to

give the shoes a special character.

棉帆布鞋面可在装配前进行砂洗，以赋予鞋子特殊的风格。

3. These linked and meandering loops can be easily stretched in different directions giving knit fabrics more elasticity than woven fabrics.

这些弯曲交织的线环可以很容易地向不同的方向拉伸，使针织物比机织物更有弹性。

4. Common knit types are jersey, interlock, double knit, and ribbed.

常见的针织类型有平纹、连锁、双面和罗纹。

5. It's the sizing and backing treatments that give the fabric the toughness and body to make it useful.

正是上浆和背衬处理赋予了织物韧性和质感，使其变得有用。

6. Nylex and Visa, or Visa Terry, are knit products where one side is brushed to tease up the soft fibers.

尼莱克斯和维萨，或维萨毛圈布都是针织产品，其中一侧经过了梳理以呈现出柔软的绒面。

词 汇

thread size：纱线尺寸

denier：旦尼尔（单位）

colorfastness：色牢度

warp：经纱

plain：平纹

satin：缎纹

doddy：短绒

symmetric：对称

interlock：连锁

coating：涂层，涂饰

wrinkle：起皱，皱折

sizing：涂胶，上浆

stretchability：可延伸性

finishing：整理，涂饰

weft：纬纱

twill：斜纹

basket：网眼

ripstop：格子

jersey：平针织物，绒布

double knit：双面针织物

brushing：涂刷，刷浆

tricot：经编针织物

Lesson 3　Shoe Leather

There are a variety of leather types, prices, and features available. From fashion shoes made with white **nubuck** to black blood-proof combat boots, leather is an amazing material for shoes.

Full grain aniline: A leather which has kept its full grain, has a naturally textured full pored surface.[1] The deep **aniline** coloring is achieved with dyes. It may or may not have a thin transparent finishing coat. This leather is of the highest quality and is the most expensive. Only flawless skins undergo such a treatment. Not well protected, these leathers darken well with age.[2]

Full grain pigmented (also called **top grain**): This shoe leather has retained intact its full grain and received a penetrating dye: a colored opaque finishing which gives a unified appearance, hides small defects, and protects the leather. Only high-quality raw hides are used. This treatment makes leather pleasing to look at and to touch and makes it resistant. It's the best compromise between esthetics and resistance.

Corrected and pigmented grain: To smooth out the leather surface and to hide its imperfections, wrinkles, and scratches... the skin is slightly embossed and a tinted grain film is applied to its surface. These leathers are not of the highest quality and they have a slightly artificial appearance. The surface coating, however, is quite resistant to heavy wear. This type of shoe leather is often used in **service style boots**. The leather will be waterproofed or even blood proofed depending on the requirements. You will find this leather used to make **steel toe** boots for soldiers, police, and medics.

Nubuck: The velvety appearance of this leather is obtained by a light pouncing (or buffing) of the skin, thus highlighting the grain and the **pores** of the leather.[3] To obtain a good-looking nubuck, a quality skin is generally used. It's a pleasant material, soft to the touch with velvety colors. However, nubuck is fragile and requires careful maintenance. A waterproofing treatment is mandatory. It remains a costly material. Lighter colors require higher-quality raw materials.

Crazy horse: To make this style of leather, a lower-quality full grain hide may be brushed to remove just the top surface. The leather is then treated with a heavy, waxy, and oily compound that will darken the leather. This is the rough and rugged style of leather—you may see scratches, bug bites, scars and **fat wrinkles** on the surface but that's okay, it's crazy horse. This leather will show color changes when flexed.

Split or **suede**: The leather surface remains after the top grain is removed. Suede is a very common shoe material. High-quality short nap suede can almost have the appearance of nubuck leather. Lower quality suede may be a hairy, cardboard-like, dusty off-color mess. Available in a rainbow of colors, quality suede is a stable material for casual, skate, and vulcanized shoes.

Action or coated leather: This leather product starts out as medium to low-quality split leather. The surface may be pressed or **rolled** smooth. The hide is then laminated to a thin film of flexible stretchable PU or PVC.[4] This material is cheap, durable, looks good and comes in many colors and finishes. This material is also easy for the factory to cut, there are typically very few scratches or scars showing that must be avoided. Action leather can look like full grain, nubuck and glossy **patent leather**.

重点及难点句

1. A leather which has kept its full grain, has a naturally textured full pored surface.
保持纹理的全粒面皮革具有自然纹理的多孔表面。

2. Not well protected, these leathers darken well with age.
如果保养不当，这些皮革会随着时间的推移而变暗。

3. The velvety appearance of this leather is obtained by a light pouncing (or buffing) of the skin, thus highlighting the grain and the pores of the leather.

这种皮革的天鹅绒般的外观是通过轻磨皮革获得的，从而突显皮革的纹理和毛孔。

4. The surface may be pressed or rolled smooth. The hide is then laminated to a thin film of flexible stretchable PU or PVC.

皮革通过辊压以获得光滑的表面，然后与一层柔韧可延伸的 PU 或 PVC 薄膜复合。

词 汇

nubuck：正绒面革

full grain aniline：全粒面苯胺革

aniline：苯胺

full grain pigmented (top grain)：全粒面染色革（头层革）

corrected and pigmented grain：修面染色革

service style boots：军靴，制式靴　　steel toe：钢包头

pore：小孔，毛孔　　crazy horse：疯马皮

fat wrinkle：肥皱　　split：剖层革

suede：反绒革

action or coated leather：二层革，移膜革

roll：辊压，辊筒　　patent leather：漆革，漆皮

CHAPTER 2

Materials Used for the Bottom

Lesson 1　Outsole

The sole is the most important part of a shoe. It consists of three major parts, insole, midsole and outsole. The footwear soling materials are normally classified as (1) Natural materials and (2) Synthetic materials. **Textiles**, leather, rubber and cork are the commonly used natural materials and **thermoplastic**, polyurethane and so on are the synthetic soling materials.

The outsole is the layer of the sole that touches the ground. Due to the amount of stress and general wear, this part is usually made of very rugged material. It is also crucial that it delivers enough **friction** with the floor to keep the wearer from slipping. The outsole is commonly rubber but can be high-density PU or EVA foam. Dress shoes may have leather bottoms.

1. Leather

Every respectable dress shoe brand will use leather or rubber soles, or a combination of both. Leather soles are the most classic type of shoe soles. You will frequently find them in formal shoes and boots and come in different forms. It would rarely make sense to put a leather sole on chunky, very casual, heavy **terrain boots**.[1] Or to put a formal black oxford on a massive commando rubber sole. Durability, grip, appearance and comfort are some of the things you would need to consider.

Leather soles are made from **vegetable-tanned leather** and often oak bark tanned. Before tanning the hide is cut into various sections: a pair of bends, a pair of bellies, the shoulder and cheeks[2]; these are tanned separately and differently according to the purpose for which they are the most suitable. The bends are used for soles and top pieces (the part of the heel which rests on the ground); the bellies for insoles, **throughs**, and **lifts** (heel); the shoulders for insoles, some **light soles**, and welts.

2. Rubber

Due to their overall durability and improved aesthetics, it is no longer a taboo to have rubber soles even with formal shoes. It can last years on the toughest of terrain. It is water-resistant as well. Just don't try to dry it too fast after getting it wet or it will crack.

There are a few different types of rubber soles such as **STUDDED RUBBER SOLES**: This is a rubber sole with studs on the front part of the shoe. Studded rubber soles are more waterproof, and durable but at the same time add weight to the shoe and are also thicker. **COMMANDO SOLES**: They have some **treads and medallion stars** in the center to provide traction and shock absorption. Generally, these are heavier and give a rather casual heavy look to shoes. However, in modern times there is a current trend when commando soles are applied to a tasteful last and model it can look very good. **RIDGEWAY SOLES**: Ridgeway is something between commando and studded soles. Not as thick as the commando but a little more than studded with treads and excellent grip and a rather nice profile. It's excellent and quite grippy without compromising the aesthetics. **CITY RUBBER SOLES**: It's a type of rubber sole that mimics the leather version with very discrete ridges or markings. This allows for a sleeker sole which can maintain the formality of the shoe. It's comfy, lighter and excellent for rainy environments but the grip can vary. **CREPE SOLES**: This natural rubber sole has a yellowish tint and is extremely comfortable and light with the disadvantage of quick wear. WEDGE SOLES: **Wedge soles** run the entire length of the shoe.[3] With a complete surface area, they tend to be comfortable and their treads can give good traction. However, they are extremely casual and suitable for a very specific type of shoe which is usually boots. Very popular on **moc-toes**.

3. PU or polyurethane

Polyurethanes are either polyester or polyether-based depending on the physical and mechanical properties required and can be injected or pour-molded.[4]

Polyurethanes can be made light, tough, comfortable, flexible, insulating, waterproof, slip-resistant, hardwearing and shock absorbent as required, simply by varying the formulation. They can have an almost endless variety of shapes, surface textures and colors and incorporate voids, inserts or dual densities for extra comfort and support.[5] Using polyurethanes, the manufacturer **is well placed to** address special needs such as anti-static properties, improved abrasion resistance, low-temperature flexibility and resistance to hydrolysis, microbial action and **ultraviolet yellowing**.[6]

Thermoplastic, rather than microcellular polyurethanes, offers footwear manufacturers a material with high impact resistance. This is especially suitable for applications such as sports plates for soccer, golf and baseball shoes; skates and ski boots; **top pieces** for lady's fashion shoes; and molded heels for women's shoes. Many women's fashion shoes high-heel outsoles are made from high-density PU.

4. PU/rubber system

Rubber and polyurethane (PU) is a perfect material combination for shoe soles with a wide range of best properties. The compact rubber outsole stands for high wear resistance. The foamed, lightweight PU midsole offers a pleasant wearing comfort. Combining the benefits of polyurethane and rubber, the PU/rubber system provides the benefits of polyurethane and rubber, including heat and slip resistance, shock absorption and abrasion resistance. The production of two-layer soles with a rubber outsole and PU midsole becomes faster, more cost-effective and **safer in bonding** for a wider range of sole designs.

5. EVA

The EVA stands for **ethylene vinyl acetate**. This is an elastomeric polymer that makes up materials that are rubbery in both softness and flexibility. It is a plastic fabricated through the combination of ethylene vinyl acetate to produce rubber-like properties that can be used to make soles for shoes.[7] PU is somewhat heavier with a higher density. However, in footwear, it excels in the area of durability. EVA is lightweight and less dense than polyurethane but has superior shock absorption. It will, however, compress over time which means it will eventually stop providing much support after a significant amount of wear.

6. TPR

Thermoplastic Rubber (TPR) has characteristics of both rubber and plastic. This form of synthetic rubber can melt into a liquid when heated and, much like water, it becomes a solid when cooled. It is made out of the polymer SBS (Styrene-butadiene-styrene). SBS is what is known as a tertiary "**block copolymer**". This means that there are blocks of each monomer (i.e., styrene–butadiene–styrene) inside the polymer rather than a random distribution. TPR soles are useful for handling slips on surfaces that are slippery and slick. TRP soles are also typically utilized on shoes that have the purpose of taking part in outdoor activities. They are suited for the production of waterproof TPR shoes. TPR also has a rather rough texture in comparison to the common rubber sole yet it remains lightweight.

7. PVC

PVC stands for Polyvinyl Chloride. It is a plastic polymer used in the manufacture of a wide variety of products. PVC soles are for the most part done with the direct injection process[8] but they can also be fabricated as PVC foam boards. It is very **pliable**, abrasion-resistant and inexpensive. There are upsides and downsides to PVC. PVC is durable and has good insulation. It is resistant to oil and water but the texture is poor as is the anti-slip resistance.

重点及难点句

1. It would rarely make sense to put a leather sole on chunky, very casual, heavy terrain boots.

在厚重、休闲的重型远足靴上使用皮革鞋底是不合逻辑的。

2. Before tanning the hide is cut into various sections: a pair of bends, a pair of bellies, the shoulder and cheeks.

在鞣制之前将生皮切成不同的部分：两片背部、两片腹部以及肩部和头部。

3. Wedge soles run the entire length of the shoe.

楔形片底覆盖整个鞋长。

4. Polyurethanes are either polyester or polyether-based depending on the physical and mechanical properties required and can be injected or pour-molded.

根据所需的物理和机械性能，聚氨酯或为聚酯型或为聚醚型，可以注射或浇注成型。

5. They can have an almost endless variety of shapes, surface textures and colors and incorporate voids, inserts or dual densities for extra comfort and support.

它们可以有几乎无穷无尽的形状、表面纹理和颜色，并结合空腔、插板或双密度，以获得额外的舒适度和支撑力。

6. Using polyurethanes, the manufacturer is well placed to address special needs such as anti-static properties, improved abrasion resistance, low-temperature flexibility and resistance to hydrolysis, microbial action and ultraviolet yellowing.

使用聚氨酯，制造商可以很好地满足特殊需求，如抗静电、提高耐磨性、耐低温曲挠、耐水解、微生物作用和紫外线黄变等。

7. It is a plastic fabricated through the combination of ethylene vinyl acetate to produce rubber-like properties that can be used to make soles for shoes.

它是一种由乙烯和乙酸乙烯酯结合制成的塑料，与橡胶性能类似，可用于制造鞋底。

8. PVC soles are for the most part done with the direct injection process.
PVC 鞋底大部分采用直接注射工艺。

词 汇

textiles：纺织品

thermoplastic：热塑性的，热塑性塑料

friction：摩擦，摩擦力

terrain boots：远足靴

vegetable-tanned leather：植鞣革

through：中底

lift：后跟皮；插跟

light sole：轻质鞋底

studded rubber sole：有圆形凸起的橡胶底，镶钉橡胶底

commando sole：登山鞋底

treads and medallion stars：条带状和星状花纹

ridgeway sole：山脊路鞋底

city rubber sole：城市鞋底，仿皮底

crepe sole：绉胶底，生胶底

wedge sole：楔形片底（前薄后厚）

moc-toe：皱头，莫卡辛鞋头

be well placed to：处于有利地位

ultraviolet yellowing：紫外线黄变

top piece：天皮，鞋跟面皮

safe in bonding：黏合牢固

EVA (ethyl vinyl acetate)：EVA（乙烯 – 醋酸乙酯）

block copolymer：嵌段共聚物

pliable：柔韧的，柔软的

Lesson 2 Insoles

Insoles are pieces of material that are placed inside your shoes or boots. Sometimes referred to 'footbeds' or 'inner soles'. Insoles are also a way to keep shoes fresh as they can be taken out and cleaned. Insoles come in various materials which have different pros and cons. These include:

Foam insoles: The most common, cheapest and widely found material is foam. Foam is naturally shock-absorbing, rigid and always dependable. However, it can wear down quickly.

EVA (ethyl vinyl acetate) is the copolymer of ethylene and vinyl acetate and the most common midsole material for sports shoes. EVA is lightweight, durable, easy to form, and resists compression set. EVA can be hot pressed, cold pressed, die cut, injected, and machined to make midsoles or inserts.[1] Available in a wide range of densities and formulations, EVA can be soft and flexible or rock hard and stiff and be made in almost any color. It can be found in all different styles of shoes. EVA is also used to make footbeds, and **padded Stroble socks**, and is often laminated as a fabric backing.

Closed cell PU (Polyurethane): Also a common foam for shoes. PU foam is "blown" into molds. The liquid compound expands and foam air cells fill the molds. PU foams are used to make durable midsoles for hiking boots and can be made into entire sole units, **tread**, and midsole all in one.[2] Very soft PU is used for footbeds due to its resistance to **compression set.**

Memory foam insoles: These are molded perfectly to the shape of your foot, which makes them very comfortable to wear and longer lasting.

Gel insoles: These are a smart buy for heels that are too big, as they increase grip, stop slipping and relieve pain, especially in the balls of the feet.[3]

Air-cushioned insoles: These provide maximum comfort and cushioning. Air-cushioned insoles contain pressurized air bubbles inside a supportive foam sole, prolonging the life of your shoes.[4]

Leather insoles: Leather is another insole option. Cowhide leather is tough and hardwearing, providing strong arch support.

The most common reason to choose an insole is to make your shoe feel slightly smaller and fit better. But insoles can also provide:

A better fit: Standard shoe sizes can fit differently depending on where you shop, and it can often be difficult to find a half size. Sometimes the best option is to choose a shoe that is slightly too big, then add an insole to fill up a little more of the shoe for a tighter fit. The best insole for shoes that are too big depends on your personal preference. You can always layer a couple of different insoles on top of each other to pad the shoe out.

Added warmth: As well as making your shoes **fit** better, insoles can also help your feet feel warmer. Adding an extra layer that helps you to insulate your feet in colder weather.

Prevention against blisters: Insoles can also help to prevent blisters. A tighter-fitting shoe won't rub as much and cause irritation and discomfort.

Improved posture: Insoles can also be used to help correct issues with posture. These types of insoles are specialized and are known as orthotics or orthotic insoles. They can be custom-made to provide appropriate treatment for certain problems that affect standing, walking or running—providing arch support where it is needed.[5]

Other common foams for shoes

PE (Polyethylene): Expanded into sheets, PE foam is easily die-cut and laminated. Parts are then pressed into shape for internal pads and tongues. Due to its weakness in compression set, PE foam is not used underfoot, but PE foam is closed cell and waterproof.

SBR (Styrene butadiene rubber): A very soft foam, often laminated between two layers of fabric. SBR is a closed cell and is used to make parts waterproof. SBR foam is often used as a lightweight replacement for neoprene rubber but is not as stretchable as it.

Open cell PU (Polyurethane): This may be the most common foam used in sports shoe's construction. This PU foam is open cell so you must be careful that it does not absorb the glue. Thin layers of PU are laminated to fabric to provide a backing substance. PU foam is also used to make **tongue foam** and **collar foam**. Due to its softness, open-cell PU foam cannot be used underfoot.

Latex rubber foam: Latex foam is easily formed into complex shapes in open-top molds. Latex is used for upper padding but not as a midsole material. High-density latex foam sheets are often used to make die-cut footbeds.

重点及难点句

1. EVA can be hot pressed, cold pressed, die cut, injected, and machined to make midsoles or inserts.

EVA 可以进行热压、冷压、模切、注射，并被加工成中底或嵌入件。

2. PU foams are used to make durable midsoles for hiking boots and can be made into entire sole units, tread, and midsole all in one.

PU 泡沫可用于制作耐用的登山靴中底，也可制成集外底面和中底于一体的整体鞋底。

3. These are a smart buy for heels that are too big, as they increase grip, stop slipping and relieve pain, especially in the balls of the feet.

对于太高的高跟鞋来说，这（选用凝胶鞋垫）是一个明智的选择，因为它们增加抱脚力、防（止脚在鞋腔内）滑（动）并且缓解尤其是脚掌部位的疼痛。

4. Air-cushioned insoles contain pressurized air bubbles inside a supportive foam sole, prolonging the life of your shoes.

位于支撑型外底中的气垫中底含有加压气泡，用以延长鞋的寿命。

5. They can be custom-made to provide appropriate treatment for certain problems that affect standing, walking or running—providing arch support where it is needed.

定制的鞋垫可对影响站立、行走或跑步的某些问题提供适当的治疗，即在需要的地方提供足弓支撑。

词 汇

foam insole：发泡（材质）鞋垫

padded Stroble sock：加衬锁缝鞋套（闯楦成型法）

tread：鞋底，外底面

compression set：压缩变形

air-cushioned insole：气垫鞋垫

fit：合脚

tongue foam：鞋舌泡棉

collar foam：鞋口泡棉，鞋领口泡棉

Lesson 3   Adhesives in Shoemaking

The main use of adhesives in shoemaking is to attach the sole to the upper. During 1950-1965, **polychloroprene adhesives** exhibited excellent **coalescing properties** with the main shoemaking materials of the day, leather and rubber, and also proved capable of heat reactivation days or even weeks after application before the actual bonding operation with a short press time.[1]

Then during the mid-1960s, the footwear industry started taking a close interest in plasticized PVC soles and upper materials. Polychloroprene adhesives proved inadequate but fortuitously **polyurethane adhesives** were available. These exhibited excellent adhesion to PVC as well as offering inherent resistance to grease and **plasticizers**.[2] Standard polyurethane adhesives were not so successful with thermoplastic and vulcanized rubbers, until SATRA developed a surface priming treatment, which they called Satreat.

From the mid-1980s, **water-based** and hot-melt adhesives were under close development and progress continued through the 1990s in response to increasingly stringent limits on **VOC** emissions to the atmosphere and health concerns for operatives.

Water-based systems for both polyurethane and polychloroprene adhesives are widely available today and are used in a very similar manner to solvent-based systems. Developments have overcome early difficulties such as a lack of **initial bond strength** that limited the new system to softer sole materials such as thermoplastic rubber. Another problem was long drying times that have been solved by improved water evaporation techniques while the use of stainless steel has obviated plant corrosion problems.[3] For manual sole bonding, the adhesive dries fast enough for normal throughput speeds but automated applications can easily be integrated into the work process with a drying tunnel incorporated into the line.

The other environment-friendly adhesive development is the hot melt which has the advantage of 100 percent **solids content**. However, conventional hot melt adhesives, used for the lasting process, have relatively high **viscosity**, leading to poor wetting and poor **penetration** of leather and other fibrous materials. The problem has been overcome in reactive polyurethane hot melts by applying a partially cured polymer as a low-viscosity melt.[4] The adhesive is then further cured by moisture or heat before either direct bonding or heat reactivation. The final cure takes place after bonding.

A fascinating development in adhesive technology leads to simplified bonding procedures and a stronger bond using **self-stratifying polymer** technology developed. Instead of having to apply, for example, a **primer coat** to a surface and then a **topcoat**, only one coat of self-stratifying paint is applied.[5] This then separates into two layers to form a primer layer and a topcoat. This approach has been developed for footwear adhesives and it is now possible to avoid using a primer on the sole of some materials. Indeed, for some materials, it is even possible to apply adhesive only to the upper, saving time and money as well as producing stronger bonds.

Adhesive-coated materials are used widely in shoemaking in rolls, sheets or as pre-cut or molded components.[6] EVA adhesive is employed in many of these applications. One of the most interesting adhesive-coated components to be introduced in recent years has been the patented insoling material. Its use obviates the conventional insole and conventional cement lasting since **thermo adhesive** is impregnated on one side of the insole that is reactivated at lasting by a short blast of air.[7] Lasting itself takes about four seconds.

重点及难点句

1. ... proved capable of heat reactivation days or even weeks after application before the actual bonding operation with a short press time.

……刷胶后几天甚至几周内，在短时间压合的实际黏合操作之前，能够重新热活化。

2. These exhibited excellent adhesion to PVC as well as offering inherent resistance to grease and plasticizers.

这些产品（聚氨酯胶黏剂）对聚氯乙烯表现出优异的黏合性，并具有内在的抗油脂和增塑剂特性。

3. Another problem was long drying times that have been solved by improved water evaporation techniques while the use of stainless steel has obviated plant

corrosion problems.

另一个问题是干燥时间长，这已经通过改进的水蒸发技术得到了解决，而不锈钢的使用避免了设备腐蚀。

4. The problem has been overcome in reactive polyurethane hot melts by applying a partially cured polymer as a low-viscosity melt.

在反应性聚氨酯热熔胶中，通过使用部分交联的聚合物作为低黏度熔体，已经克服了这个问题。

5. Instead of having to apply, for example, a primer coat to a surface and then a topcoat, only one coat of self-stratifying paint is applied.

例如，不必在表面涂底胶，然后再涂面胶，只需涂一层自分层胶料。

6. Adhesive-coated materials are used widely in shoemaking in rolls, sheets or as pre-cut or molded components.

制鞋中广泛使用的背胶材料（材料背面已涂胶）是成卷的、成片的，预切割或模压成型的部件。

7. Its use obviates the conventional insole and conventional cement lasting since thermo adhesive is impregnated on one side of the insole that is reactivated at lasting by a short blast of air.

它的使用避免了采用传统内底和传统的刷胶绷帮法，因为热熔胶已被浸渍在内底的一侧，可在绷帮时通过短促的气流重新活化。

词 汇

polychloroprene adhesive：氯丁橡胶胶黏剂，氯丁胶

coalescing property：黏合性能

polyurethane adhesive：聚氨酯胶黏剂

plasticizer：增塑剂

water-based：水基型的，水性的

volatile organic compounds（VOC）：挥发性有机化合物

initial bond strength：初始黏合强度

solids content：固（体）含量

viscosity：黏性，黏度

penetration：渗透性

self-stratifying polymer：自分层聚合物

primer coat：底涂层

topcoat：顶涂层

thermo adhesive：热固性胶黏剂

CHAPTER 3

Quality Control of Shoes

Lesson 1　How to Classify Quality Defects of Shoes

Shoes aren't made to last forever. Even well-made pairs will eventually show signs of wear and tear with constant use.[1] But that doesn't always mean the customer is blamed for any wear and tear or other problems with shoes they might buy. Often, it's the manufacturer who is responsible for managing quality defects of shoes so that customers are satisfied with the finished product.[2]

Whether you're an importer of shoes who wants to improve quality at your factory in Indonesia or an end-consumer picking out a high-quality pair from a retail store, it helps to know what to look for.

1. How to classify quality defects of shoes

Before we get into the top common quality defects in shoes, it's helpful to understand how quality control professionals typically classify and sort defects by severity. Quality defects for shoes are generally sorted into one of the following three categories:

Defects that fail to meet mandatory regulations or pose a hazard to the user are typically classified as "**critical defects**".[3]

Defects that render the footwear unacceptable by affecting appearance, durability and **salability** are called "**major defects**". A major defect found in a shoe is likely to result in a customer returning the pair.

Defects that are lower than the desired quality standard but are not likely to result in customer dissatisfaction, product return or complaints are called "**minor defects**".

2. Shoes zoning for defect classification

Figure 1   Shoe zoning for detect classification

Most shoes can be divided into two zones, "Zone 1" and "Zone 2". Zone 1 is generally the most important area in terms of visual appeal because it's the most apparent area to the customer or wearer.[4] Zone 2 is less important because it includes areas of the shoe that are less obvious to the customer or wearer.

Any defects found in Zone 1 that aren't critical are more noticeable and more likely to be considered as major defects.[5] The same defect found in Zone 2 is more likely to be classified as a minor defect.

重点及难点句

1. Even well-made pairs will eventually show signs of wear and tear with constant use.

即使是做工精良的鞋子，在长期使用后最终也会出现磨损的迹象。

2. Often, it's the manufacturer who is responsible for managing quality defects of shoes so that customers are satisfied with the finished product.

通常，制造商负责管理鞋子的质量缺陷，以便使客户对成品满意。

3. Defects that fail to meet mandatory regulations or pose a hazard to the user are typically classified as "critical defects".

不符合强制性规定或对用户构成危险的缺陷通常被归类为"重要缺陷"。

4. Most shoes can be divided into two zones, "Zone 1" and "Zone 2". Zone 1 is generally the most important area in terms of visual appeal because it's the most apparent area to the customer or wearer.

大多数鞋子可以分为两个区域，"一区"和"二区"。就视觉吸引力而言，一区通常是最重要的区域，因为它对顾客或穿着者来说是最明显的区域。

5. Any defects found in Zone 1 that aren't critical are more noticeable and more likely to be considered as major defects.

在一区发现的任何不重要的缺陷都更加明显，并且更有可能被认为是主要缺陷。

词 汇

critical defect：严重缺陷
major defect：主要缺陷
shoes zoning：鞋子分区

salability：畅销，出售
minor defect：轻微缺陷

Lesson 2   Common Quality Defects in Shoes

Just as garment manufacturers typically use a lot of cutting and sewing during production, those that manufacture footwear often use some similar production processes. Although shoes are typically made with more robust materials, such as firm leather, rubber and plastics, they're vulnerable to similar product defects. Still, there are some common quality defects that are unique to shoes.[1]

1. Excess glue, wax or oil

Excess glue, wax or oil marks are the most common quality defects found in shoes. These types of residue are especially common because:

Most of the shoe production **facilities** use adhesives and other chemicals during production.

Factory workers are less likely to protect against or remedy issues like excess glue, wax or oil when rushing to complete an order, which is often the case for most factories.[2]

Glue, wax or oil residue can often be cleaned off without difficulty. But if this defect is evident in the finished product, it can be very unsightly to customers and possibly renders a shoe unsellable. Excess glue, wax and oil marks are often found in both zone 1 and 2 areas.

Preventing excess glue, wax and oil:

Shoes are often left with material residue because of the chemicals or adhesives they're exposed to during manufacturing. But there are some simple ways of preventing this quality defect from remaining on the finished goods, namely:

Make sure factory workers aren't using too much glue or other chemicals during production.

Any excess material left on the shoes after production should be wiped away prior to packaging.

There's no reason why you can't greatly reduce or eliminate the occurrence of this quality defect because it's relatively easy to remedy after the fact.[3]

2. Degumming or weak cementing

Usually evident on shoes with rubber soles, such as sneakers, **degumming** or **weak cementing** happens when there is an insufficient adhesive used when applying the sole to the upper part of the shoe.[4] This defect is usually found between the join lines of zones 1 and 2. Generally, however, the problem is considered a Zone 1 defect and a major defect.

Preventing degumming or weak cementing:

It's important to make sure that the factory manufacturing the footwear is using the correct type of adhesive. However, the more common cause for weak cementing or degumming is not enough adhesive applied between shoe components. Make sure workers are using enough adhesive but not too much. You'll often find excess adhesive apparent around the **seal** if too much adhesive was used during binding.

3. Abrasion marks

Abrasion marks are a type of quality defect in shoes that normally appear in Zone 1. These are often considered more serious when found on leather shoes or those with a glossy surface because they're more obvious.

Abrasion marks are usually caused by poor handling by factory workers during the production process.

Addressing abrasion marks found on shoes:

If you're an importer finding abrasion marks on a significant number of pieces in an order, there are a few factors that you should investigate, such as:

Are factory workers handling the product roughly? Are they wearing gloves while working?

Are the shoes subject to a lot of unnecessary moving between workstations?

Does the packaging provide enough protection to prevent abrasions during transit?

Abrasions are not always easy to spot. That's why having a **golden sample** with you or a third-party inspection company while checking shoes is extremely useful. It can provide a helpful contrast between what is acceptable and what is not.

4. Asymmetry in shoes

Asymmetry can be an issue where different components of shoes do not line up as they should. Some examples of asymmetry commonly found on a shoe or pair of shoes are:

Where the sole of the shoe does not line up with the body from a front, rear or side view;[5]

Where certain parts of the shoe aren't straight, such as the tongue;

Where part of one shoe is higher or lower than the same part of its counterpart in a pair[6] (often called "hi-low").

Asymmetry in shoes is often related to issues with the cutting or fitting of the components.

Finding asymmetry in shoes:

Issues with shoe asymmetry should be addressed with the manufacturer. The best way to find asymmetry in shoes is to place them side by side or back to back. Essentially, you need to determine whether or not the shoes reflect each other in terms of height, width, color, and so on.[7] For example, if you place shoes back to back, you need to observe that the height of the heel is the same on each one, otherwise you have a defect.

5. Incorrect sizing

Most standard shoe sizing tools just last for a while.[8] It's possible, however, that due to an error during the production process, the shoe was labelled and packaged in the incorrect size. This should be considered a major defect because shoes that don't match specifications for size are likely going to be unsellable.

Preventing incorrect shoe sizing:

It's unusual to find shoe sizing inconsistent due to production processes. Incorrect sizing is usually a result of the way the finished shoes are **sorted** and packaged. Naturally, shoe sizing discrepancies are more common in factories that are disorganized. And often all that's needed to prevent incorrect sizing is better procedures for handling, packaging and storing the finished products. Taking a look at a factory's warehouse and packaging areas can tell you a lot about the likelihood of incorrect sizing.

6. Protruding nails or sharp point

A protruding nail or sharp point is among the least common quality defects in shoes on our list. As with the previous defects, this is not a visual defect, so it does not apply to zones 1 or 2.

Some shoes, especially those made of leather, have nails to bind the sole of the shoes to its upper part during production. If these nails are not properly pressed down, they can protrude into the insole, creating a sharp point that could hurt the consumer. In other cases, a needle might be mistakenly left in a shoe.

Preventing nails or sharp points in shoes:

As with garment factories, footwear factories should and generally be equipped with metal detectors. After production is completed, workers set the units on a belt that runs through the machine to look for needles or other sharps that might have been left by mistake.

Factories should also have QC staff examining shoes off the line for signs of sharp points that could be harmful to the end consumer. If sharp points are found in your order or shoes, the factory needs to investigate which process is responsible and eliminate the cause.

Shoes can be bought to make a fashion statement, or simply provide utility and comfort. Either way, most shoes are manufactured with some level of quality in mind. As a customer or manufacturer, it helps to look out for some of the common quality defects in shoes that could affect you or your business.

With awareness of these defects, you can safeguard yourself against shoddy quality. Ultimately, inspecting your shoes at the source before they ship is the best way to ensure you're getting the quality you paid for.

重点及难点句

1. ... , they're vulnerable to similar product defects. Still, there are some common quality defects that are unique to shoes.

……，它们很容易出现类似的产品缺陷。尽管如此，鞋子还是有一些独特的质量缺陷。

2. Factory workers are less likely to protect against or remedy issues like excess glue, wax or oil when rushing to complete an order, which is often the case for most factories.

在匆忙完成订单时，工厂工人不太可能防止或解决胶水、蜡或油等过量使用的问题，且大多数工厂都经常出现这种情况。

3. There's no reason why you can't greatly reduce or eliminate the occurrence of this quality defect because it's relatively easy to remedy after the fact.

没有理由不能大大减少或消除这种质量缺陷的发生，因为事后补救相对容易。

4. Usually evident on shoes with rubber soles, such as sneakers, degumming or weak cementing happens when there is an insufficient adhesive used when applying the sole to the upper part of the shoe.

当把鞋底与鞋帮黏合时，如果没有使用足够的黏合剂，脱胶或黏合不良就会发生，通常在橡胶底的鞋子上很明显，如运动鞋。

5. ... where the sole of the shoe does not line up with the body from a front, rear or side view.

……从前面、后面或侧面看，鞋底与鞋身不齐。

6. ... where part of one shoe is higher or lower than the same part of its counterpart in a pair.

……一只鞋的某一部件比另一只鞋的同一部件高或低。

7. Essentially, you need to determine whether or not the shoes reflect each other in terms of height, width, color, and so on.

从本质上讲，你需要确定（同双的两只）鞋子是否在高度、宽度、颜色等方面相一致。

8. Most standard shoe sizing tools just last for a while.

大多数标准的鞋测量工具只能用一段时间。

词　汇

facility：设备，工具，设施

weak cementing：黏合不良

abrasion mark：擦痕，磨痕

incorrect sizing：错码

protruding nail：露钉，突出的钉子

degumming：脱胶

seal：黏合处，接缝处

golden sample：标（准）样（品）

sort：（鞋的）分拣

Lesson 3   Key Areas of Quality Control Inspection for Footwear

A shoe often looks like a simple, cohesive unit once it's a finished product, but footwear manufacturing requires careful workmanship and assembly of multiple parts. And footwear manufacturing is still largely a labor-intensive process. So human error can lead to numerous defects in the finished goods. If you intend to deliver shoes that meet your customers' standards for comfort, durability, style and quality, pre-shipment footwear inspection is essential.[1]

Careful measurement and **fitting check** of the shoe's insole length, insole width, **toe lift**, heel height and back height for proper fitting, symmetry and sizing requirements are needed.

Checking for issues in stitching includes conducting seam strength tests and checks for fraying, **skipped stitches, feed abrasion marks, open stitches, crooked stitches and puckered seams**.[2]

Eyelet inspection includes checks of pairing, spacing, sharp parts, deformation and looseness.[3]

Heel inspection includs checks for denting, **webbing** and proper attachment by nails, screws or staples.[4]

On-site testing includes marking and vulcanization tests for rubber outsoles, function and fatigue tests on **closures**, bonding tests, flex tests, **rocking tests**, fabric testing and verification of special features like waterproofing and stain resistance.[5]

Checking conformance to legal requirements for footwear includes protective footwear standards, children's footwear requirements and **labeling requirements for country of origin** and material composition.

Packaging inspection includes check of polybags, barcodes and carton assortment for conformance to requirements, as well as checks of retail packaging, such as shoe position within retail shoebox, proper sizing stickers, **tissue paper** or paper card forms and silica gels.

Coordinating testing by an accredited laboratory to test shoe properties like slip resistance, sole wearing resistance and shoelace durability, as well as to identify issues in fabrics such as harmful substances, colorfastness, flammability, fiber analysis, light exposure and wet and dry cleaning properties.[6]

Shoe inspection quality "A", "B", "C" Grades

A-grade shoes: Shoes without any functional defects or **cosmetic defects** that will impair the marketability of the shoe are A-grade. These are high-quality shoes that look good and fit correctly. An A-grade must follow the production specifications and match the approved **confirmation sample**.

B-grade shoes: Shoes without any major functional defects and which will not cause injury to the person wearing the shoes are B-grade. B-grade shoes may have cosmetic defects, production mistakes, or workmanship issues that cannot be properly repaired. These B-grade shoes will be discounted and/or diverted to markets more tolerant of cosmetic defects.

C-grade shoes: C-grade shoes have major functional defects that could cause injury to the wearer or major cosmetic defects that cannot be repaired. Shoes are also considered C-grade if they have poor workmanship or materials defects that could shorten the normal life expectancy of the shoe, or damage the company's reputation. These shoes should be destroyed.

重点及难点句

1. If you intend to deliver shoes that meet your customers' standards for comfort, durability, style and quality, pre-shipment footwear inspection is essential.

如果您打算交付符合客户对舒适度、耐久性、款式和质量标准的鞋子，装运前鞋类检查是必不可少的。

2. Checking for issues in stitching includes conducting seam strength tests and checks for fraying, skipped stitches, feed abrasion marks, open stitches, crooked stitches and puckered seams.

缝制检查题包括测试缝合强度和检查缝线发毛（磨损）、跳线、送料磨损痕迹、浮线、缝线不直及鼓包。

3. Eyelet inspection includes checks of pairing, spacing, sharp parts, deformation and looseness.

孔眼检查包括配对、间距、尖锐部分、变形、松动。

4. Heel inspection includes checks for denting, webbing and proper attachment by nails, screws or staples.

鞋跟检查包括（跟体表面）凹陷、织带和用钉子、螺钉或卡钉的钉跟牢度。

5. On-site testing includes marking and vulcanization tests for rubber outsoles, function and fatigue tests on closures, bonding tests, flex tests, rocking tests, fabric testing and verification of special features like waterproofing and stain resistance.

现场测试包括橡胶外底的留痕和硫化测试，闭合件的功能和疲劳测试，黏接测试，弯曲测试、摇晃测试，织物测试以及防水、耐污等特性的验证。

6. Coordinating testing by an accredited laboratory to test shoe properties like slip resistance, sole wearing resistance and shoelace durability, as well as to identify issues in fabrics such as harmful substances, colorfastness, flammability, fiber analysis, light exposure and wet and dry cleaning properties.

由认证的实验室协调测试，主要测试鞋的性能，如防滑性、鞋底耐磨性和鞋带耐久性，以及确定织物的有害物质、色牢度、可燃性、纤维分析、耐光以及湿洗和干洗性能。

词　汇

fitting check：合脚性检验　　　　　toe lift：（鞋的）前跷

skipped stitch：跳线　　　　　　　feed abrasion mark：进料磨损痕迹

open stitch：浮线　　　　　　　　crooked stitch：缝线不直

puckered seam：鼓包，缝合不平伏　webbing：织带

closure：闭合件，闭合方式

rocking test：摇晃测试（带跟鞋的稳定性测试）

labeling requirements for country of origin：原产国标签要求

tissue paper：防潮纸　　　　　　　cosmetic defect：外观缺陷

confirmation sample：确认样（品）

CHAPTER 4

Quality Inspection of Shoes

Lesson 1   Testing Items on Whole Shoes

Due to the complexity of global supply chains, there are many possibilities for safety, manufacturing, and chemical issues to arise in footwear and footwear component production. Therefore, it's important to detect and identify any problems that could prevent or restrict your product from being retailed.

The shoes are tested in different items depending on what they are used for.

1. Slip resistance

Slips, trips and falls are the biggest causes of footwear-related accidents. The risk of compensation claims for injuries due to slipping is a major concern for footwear companies.[1] Ensuring all footwear has an adequate level of slip resistance is essential. Slip resistance and traction are extremely important in footwear designed for performance in more difficult terrain whether this is a cross-trainer, hiking or mountain boot, or a boot intended for military application. It is important that the sole is constructed to meet the challenges of a range of conditions. Good slip resistance is also an essential feature of footwear in industrial and other work environments.

2. Comfort

Comfort sells footwear. If the customer does not feel comfortable in the shoes at the point of sale, he or she is less likely to make a purchase. Any special feature that makes one brand more comfortable than another may help to **seal the deal**. Comfort plays a significant role in **performance footwear** where the wearer will have high expectations.

3. Chemicals

The list of chemicals that come under legislative restrictions is growing continuously.[2] Requirements such as REACH in Europe and Proposition 65 in the US mean that footwear must be assessed for compliance.[3]

4. Durability (especially for safety/performance footwear)

Wearers want to get the maximum life from a pair of shoes. Safety footwear should withstand many types of environments and **premature wear** could compromise protective features. Sole durability is critical in high-performance footwear.

5. Breathability (especially for safety/performance footwear)

High **breathability** can offer added value to everyday footwear. If footwear is to be worn in extreme conditions where heat and cold can vary considerably it is important to keep the wearer's feet as dry as possible for comfort and to maintain foot health. In cold environments the foot can feel even colder if thermal conductivity increases due to damp or wet conditions inside the footwear. Safety footwear is often worn for extended periods of time; therefore, breathability will be an important factor for user's comfort. Comfort itself can be a factor in helping to maintain safety by minimizing foot fatigue.

6. Fastenings

The role of good quality fastening systems, from traditional laces to elastics and **touch and close fasteners** should not be underestimated. As well as maintaining good fit they keep the shoe on the foot and also play an important role in the footwear's appearance.

7. Fitting

Shoes might be on the foot all day. Not only will a good-fitting shoe influence customer decision-making at the point of sale, but it will also assist in keeping the wearer comfortable and, hence, increase the chance of repeat sales.[4] Performance footwear must fit correctly to aid foot comfort and help minimize the risk of blisters or other injuries.

8. Heel strength (especially for fashion footwear)

Badly designed heels can crack or break as a result of impact or fatigue during everyday walking and general use. Heel failures can result in serious injury to the wearer.

9. Heel attachment (especially for fashion footwear)

High-heeled shoes need to be engineered using the correct components to ensure the heel is attached securely to the backpart.

10. Backpart stability (especially for fashion footwear)

The backpart of any high-heeled shoe should be rigid enough to ensure the wearer has the maximum stability.

11. Trim attachment (especially for fashion/children's footwear)

Trims and accessories are popular in the fashion footwear market. Making sure they are attached correctly will avoid customer's disappointment and damage brand reputation. A loose **trim** on an infant's shoe is a potential tripping or choking hazard.[5]

12. Strap strength (especially for fashion footwear)

Straps and fastenings play an important part in ensuring a good fit and, of course, keeping the shoe secure on the foot. Strap failure can result in injuries to the wearer.

13. Toe cap strength (especially for safety footwear)

Toe cap strength which is under impact and compression is the key feature of the most protective footwear.

14. Thermal rating (especially for safety/performance footwear)

High **thermal insulation performance** is particularly important for the comfort and safety of wearers in cold conditions. Conversely, low insulation is desirable when trying to keep feet cool in hot environments. Safety footwear is often used in extreme environments so the wearer's feet need to be kept at a comfortable temperature.

15. Ergonomics (especially for safety footwear)

To be practical and, of course, safe, protective footwear must be ergonomically acceptable. Anything that makes it difficult for the wearer to carry out their job properly will inhibit their safety and might even discourage wear altogether. International standards usually contain a basic ergonomic assessment.

16. Flexibility (especially for performance footwear)

A degree of stiffness may be beneficial in certain types of performance footwear, for example, in walking boots where extra support and protection are required. If footwear is too rigid, however, the support becomes counter-productive.[6]

17. Shock absorption and ground insulation (especially for performance footwear)

Although the human body can be very effective in absorbing shock when jumping, running or walking, injuries can still occur through impact loading to the foot. A number of health problems have been linked to **one-off shock loading** or repeated shock loading. These include lower back pain, **joint disorders** and heel bone fractures.

18. Water resistance (especially for performance footwear)

Most outdoor performance footwear worn in temperate or tropical climates needs to be water resistant for extended periods. Water resistance is not just a matter of keeping the wearer's feet dry but also controlling the amount of water absorbed by the upper or other components of the footwear. Water absorption will make the footwear heavier and may affect the insulating properties or cause components such as insoles to break down.

重点及难点句

1. The risk of compensation claims for injuries due to slipping is a major concern for footwear companies.
因滑倒受伤而要求赔偿的风险是鞋类公司的一个主要担忧。

2. The list of chemicals that come under legislative restrictions is growing continuously.
受到立法限制的化学品（清单）在不断增多。

3. Requirements such as REACH in Europe and Proposition 65 in the US mean that footwear must be assessed for compliance.

欧洲的 REACH 法规和美国的 65 号提案等要求意味着必须对鞋类进行合规评估。

4. Not only will a good-fitting shoe influence customer decision-making at the point of sale, but it will also assist in keeping the wearer comfortable and, hence, increase the chance of repeat sales.

一双合脚的鞋子不仅会影响顾客在销售点的决定，它还会提供给穿着者舒适感，从而增加再次销售的可能性。

5. A loose trim on an infant's shoe is a potential tripping or choking hazard.

婴儿鞋上固定不牢的装饰件有绊倒或窒息的潜在危险。

6. If footwear is too rigid, however, the support becomes counter-productive.

然而如果鞋子太硬，这种支撑就会适得其反。

词 汇

seal the deal：达成交易

performance footwear：功能鞋

premature wear：过早磨损

breathability：透气性

touch and close fastener：粘扣

heel strength：鞋跟强度

heel attachment：装跟牢度，鞋跟附着强度

backpart stability：（鞋）后跟部位稳定性

trim：装饰件，饰扣，（制革）修边，铣削，装饰用革

thermal insulation performance：保温性能

one-off shock loading：一次性冲击载荷

joint disorder：关节紊乱

Lesson 2   Slip Resistance Testing

1. What does "slip resistant" mean?

In terms of footwear, slip resistance is the amount of resistance the sole of the shoe exerts while moving over the walking surface. Slip resistance is measured by the Coefficient of Friction (COF), which is equal to the ratio between the maximal frictional force that the sole of the shoe exerts and the force pushing the sole across the surface.[1] There is an entire system for measuring slip resistance, both in flooring and footwear.

2. How is slip resistance measured?

Some people think that a shoe is either slip-resistant or not **slip-resistant**, but slip resistance is measured on a scale based on the COF. The scale runs from 0.00, equivalent to an ice-skating rink or another extremely slippery condition, up to 1.00, which is like walking on a dry carpet. So basically, the higher the number, the greater the slip resistance.

So rather than ask whether or not a pair of shoes is slip-resistant, it is more imperative that you discover HOW slip-resistant a pair of shoes is.[2] A shoe's slip resistance rating depends on its ability to resist sliding in a given set of conditions. The better the shoe grips, the higher the rating. Not all slip-resistant footwear performs equally.

A variety of factors affect the performance of a shoe including tread pattern, contact between outsole surface and flooring, and outsole compound technology.

3. What methods exist for testing slip resistance?

The first slip resistance measurement device was invented in the 1930s and there have been nearly 100 such devices invented since. There are a variety of slip resistance test machines (English XL, Brungraber Mark II, SATRA STM 603, and others), and a wide variance in test conditions (dry, wet, contaminated with oil, soap, etc.). Some machines measure slip resistance based on how much force it takes for an outsole to begin slipping while others test based on how much force can be applied before a slip occurs.

4. What does HS Oily/Wet mean?

There are several floor variables that are tested for. Most often you will see results from three: Dry, Wet and HS Oily/Wet. By testing for all these conditions, it gives an idea of how a slip-resistant outsole will perform under those various stressful conditions in the workplace. All of these tests are performed on a square of clean Red American Olean Quarry tile.

"Dry" means that a square of dry tile is used as the walking surface.

"Wet" means that a square of tile is covered with 25mL of distilled water.

"Hi Soil Oily/Wet" means that a square of tile is covered with 25mL of distilled water and 0.2g of vegetable oil.

The most important condition to consider is HS Oily/Wet because this is akin to the toughest work environments you will face. How a shoe performs under these conditions is the true judge of a shoe's slip resistance rating.[3] A shoe can have a perfect 1.0 in dry conditions but be completely useless to you as a work shoe if it fails to protect you in greasy, oily and wet conditions.

5. Conclusion

The best thing to remember when picking out a slip-resistant shoe or safety shoe vendor is to do research. Companies should be able to provide information about the performance of their products. Just like a car maker should be able to tell you the **MPGs**, engine size, passenger limit and 0–60 **mph** of the vehicle, a shoe provider should be able to tell you how their shoe performs in HS Oily/Wet conditions. If it's not explicitly available, ask. If they can't tell you, find a company who can.

重点及难点句

1. Slip resistance is measured by the Coefficient of Friction (COF), which is equal to the ratio between the maximal frictional force that the sole of the shoe exerts and the force pushing the sole across the surface.

防滑性用摩擦系数 (COF) 来测量，它等于鞋底施加的最大摩擦力与推动鞋底沿表面前行的力之间的比例。

2. So rather than ask whether or not a pair of shoes is slip-resistant, it is more imperative that you discover HOW slip-resistant a pair of shoes is.

因此，与询问一双鞋是否防滑相比，更重要的是了解一双鞋的防滑程度。

3. How a shoe performs under these conditions is the true judge of a shoe's slip resistance rating.

鞋子在这些条件下的表现是对鞋子防滑等级的真正判断。

词 汇

slip-resistant：防滑的

miles per gallon（MPGs）：每加仑燃料所行驶的英里数

miles per hour（mph）：每小时所行驶的英里数

PART V
R&D Technology of Shoes

CHAPTER 1

Shoe CAD/CAM Technology

Lesson 1   Trends in Technology

CAD modeling background

Since shoes are 3D objects, we need to be concerned with how to specify 3D shapes, and furthermore, how to modify such shapes in order to conveniently generate precise models of the designs that we wish to produce. At the same time, the software that we use to perform the design must also provide adequate support for transforming the CAD model data into production or fabrication information.[1]

At the conceptual design stage, often the designer will sketch out the key design features for the shoe. Some sketches illustrate the styling. For example, it is customary for many brands to release **a new line of shoes** once every few months often according to season. Thus, a **summer fashion line** may have a more open upper, while a winter line may be dominated by boots. In other cases, there may only be some aspects of the shoe styling that are changed from earlier styles. For instance, an athletic shoe designer may choose to change only the upper style and materials, keeping the sole relatively unchanged.[2]

Conceptual designs traditionally are sketched physically on paper. Some modern companies use computer tools for generating sketches. The tools for this are 2D image processing software systems, for instance, Photoshop™ from Adobe Systems.

Figures 1 and 2 give two examples of conceptual designs: The first is a physical sketch highlighting some key features in the sole of a walking shoe, and the second shows a potential design for a casual shoe sketched in an image editor.

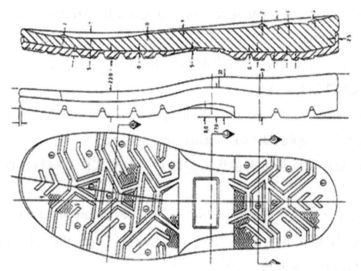

Figure 1   Sketches of a shoe sole indicating a design variation based on an earlier model

Figure 2   Sketch of a new style made by an image editor

It is important to note that image editors do not create 3D models; the outcome is merely an artist's impression (or, more correctly, the industrial designer's impression) of the potential product.[3] However, they offer several advantages including easy **archiving and retrieval**, ease of modification, powerful graphics filters including colors, texture maps, and even lighting effects to add **photorealism**.

To create the actual 3D CAD model, the designer can only use the sketch as a reference, while the entire geometry has to be recreated by using the **CAD system**.

重点及难点句

    1. At the same time, the software that we use to perform the design must also

provide adequate support for transforming the CAD model data into production or fabrication information.

同时，我们用于设计的软件也必须提供足够的支持，以将 CAD 模型数据转换为生产或制造信息。

2. For instance, an athletic shoe designer may choose to change only the upper style and materials, keeping the sole relatively unchanged.

例如，运动鞋设计师可能会选择只改变鞋面的式样和材料，保持鞋底相对不变。

3. It is important to note that image editors do not create 3D models; the outcome is merely an artist's impression (or, more correctly, the industrial designer's impression) of the potential product.

值得注意的是，图像编辑器并不创建 3D 模型；其结果仅仅是艺术家对潜在产品的印象（或者更准确地说，是工业设计师的印象）。

词 汇

a new line of shoes：新款鞋系列

summer fashion line：夏季时装系列

archiving and retrieval：存档和检索

photorealism：真实感

CAD system：计算机辅助设计系统

Lesson 2   Computer-Aided Design in Shoe Industry

Computer-aided design and manufacturing (CAD/CAM) was introduced in the shoe industry in the 1970s and focuses on designing and grading shoe upper patterns, for manufacturing of cutting dies, shoe lasts, and sole molds.[1] Initially, it was used primarily for two-dimensional(2D) pattern grading of the shoe upper. Recent traditional CAD/CAM systems used in the footwear industry today have improved with a wider range of functions including 3D footwear and decorations design; designs and manufacturing of soles; and manufacturing and machine control of shoe lasts. CAD/CAM automates routine procedures, increases speed, improves consistency, and enables design variations. CAD/CAM is used effectively in all aspects of the footwear industry as data generated at the design stage can be sent from anywhere in the world to factories for production planning and manufacturing.

The shoe upper CAD/CAM system has focused on 2D pattern generation from shoe designs, sizing and grading of upper patterns, 2D texture and logos design and engraving, optimization methods to reduce waste by properly aligning 2D patterns, machining code for cutting machines (knife or laser), and laser engraving.[2] Previously, 2D CAD was used for upper design while 3D CAD was used for sole and shoe-last design and manufacturing, but now even 3D CAD is used for upper design.

Sole CAD/CAM software is used to design and modify sole (mostly outsole), and then generate machining data to create sole molds that are used for the molding process. With computer-designed molds, the design and production cycle can be reduced, and the quality and variations in the outsole pattern have been improved. Sole design software can be used to generate molds from existing shoes using reverse engineering (scanning and modeling). Using sole CAD/CAM software complex soles can be accurately designed; the design can be easily modified based on many **geometric modeling tools**; the design changes can be visualized in real-time; the final design can now be 3D printed directly as prototypes during sample making; the 3D mold can be generated from 3D sole designs.

Shoe-last design and manufacturing CAD/CAM software are used to design and modify shoe lasts and then generate machining data to create physical shoe lasts. With computer-designed shoe lasts, the design and production cycle can be reduced, and the quality and variations of the shoe lasts have been improved. Shoe-last design and manufacturing design software can be used to generate shoe last from existing shoe lasts after **digitization** or scanning. Using shoe-last design and manufacturing CAD/CAM software complex shoe lasts can be accurately designed; the design can be easily modified based on many geometric modeling tools; the design changes can be visualized in real-time; final design can be machined using shoe-last CNC machine.[3]

The footwear manufacturing will most probably evolve into two separate directions based on footwear type: traditional footwear and 3D-printed footwear. New technology in traditional footwear manufacturing will strengthen the design (CAD), manufacturing (CAM), and engineering (CAE) components by including easy-to-use and innovative functions. CAD systems will have standalone as well as web-based systems for quick design, design changes, and design modifications. Shoe-last-based footwear design using **parametric** or point-based geometric modeling enables footwear design modifications and sizing more easily and accurately. Shoe upper design using knitting technology will improve further to have upper designs with more functions (moisture management, motion control, and functional requirements for sports).[4] Sole design will focus both on traditional techniques of making soles via molding and 3D printing technologies. Web-based footwear customization and personalization will become common as it will enable individual users to create their design using web or mobile interfaces.

重点及难点句

1. Computer-aided design and manufacturing (CAD/CAM) was introduced in the shoe industry in the 1970s and focuses on designing and grading shoe upper patterns, for manufacturing of cutting dies, shoe lasts, and sole molds.

计算机辅助设计和制造（CAD/CAM）于 20 世纪 70 年代引入鞋业，专注于鞋面样板的设计和分级，用于制造裁断刀模、鞋楦和鞋底模具。

2. The shoe upper CAD/CAM system has focused on 2D pattern generation from shoe designs, sizing and grading of upper patterns, 2D texture and logos design and

engraving, optimization methods to reduce waste by properly aligning 2D patterns, machining code for cutting machines (knife or laser), and laser engraving.

鞋面 CAD/CAM 系统侧重于从鞋类设计生成 2D 样板，帮面样板尺寸的确定和分级，二维纹理和标志的设计和雕刻，通过合理对齐二维样板来减少浪费的优化方法，切割机（刀模或激光）的机器编码，以及激光雕刻。

3. Using shoe-last design and manufacturing CAD/CAM software complex shoe lasts can be accurately designed; the design can be easily modified based on many geometric modeling tools; the design changes can be visualized in real-time; the final design can be machined using shoe-last CNC machine.

利用鞋楦设计制造 CAD/CAM 软件可以对复杂鞋楦进行精确设计；基于许多几何建模工具可以很容易地对设计进行修改；设计的改变可以实时可视化；最终的设计可采用数控刻楦机进行加工。

4. Shoe upper design using knitting technology will improve further to have upper designs with more functions (moisture management, motion control, and functional requirement for sports).

采用针织技术的鞋面设计将进一步改进鞋的功能（湿度管理、运动控制和运动的功能要求）。

词　汇

geometric modeling tools：几何建模工具
digitization：数字化
parametric：参数的，参量的

179

Lesson 3 Simulation Technologies in the Footwear Industry

The development of footwear design tools has centered its efforts during the last decades on design and manufacturing applications. Those developments have taken place under the computer-aided design (CAD) approach, and different types of CAD commercial systems depending on the footwear component focused (e.g., sole, upper) exist in the market.

The benefits provided by these CAD tools are wide and can be characterized by the type of support they give to the designer: three-dimensional (3D) design, two-dimensional (2D) patterning from 3D design, scaling to different sizes, and others. Some of these CAD tools for footwear design are able to provide a more or less realistic visual simulation of the designed footwear. This approach to the final aspect of the product is worthy from the aesthetical point of view, for example, to build virtual catalogues.[1] Concerning virtual testing of footwear, some commercial systems provide the ability to predict costs based on material consumption, through machining simulations in a CAD environment.

Based on the **CAE (computer-aided engineering)** technology, **finite elements analysis** (FEA) has become very popular in the last few years due to its versatility and accuracy. With this technology, both 2D and 3D models of the foot, sole, insole, and upper have been developed in the scientific and research fields and have been also applied in the industry to reduce the time and cost of the footwear development process.

In the last few years, a new approach merging the CAE approach with a new technological frame (CAT: computer-aided test) has been presented in the footwear industry. A virtual shoe test bed considering different levels of user-footwear functional interaction (fitting, plantar pressures, shock absorption, flexibility, friction, and **thermoregulatory** aspects) and perceptual interaction (comfort) was developed.[2]

Under that approach, the functional performance of each footwear component (sole, insole, upper, and last) is characterized through machine tests and conforming a database of footwear components ready for their selection during the design process. These test results feed a model of the footwear components interaction which is built to simulate a complete footwear functional performance. Finally, the result of the virtual analysis for the complete footwear functional performance is linked to users' comfort perception through a regression model.[3] In that sense, the designer receives worthy information on the footwear design at two levels: first, a functional performance characterization of different aspects of the user-footwear biomechanical interaction (fitting, pressure distribution, shock absorption) and, second, a prediction of the users' perception comfort when wearing such products.[4]

重点及难点句

1. This approach to the final aspect of the product is worthy from the aesthetical point of view, for example, to build virtual catalogues.

从美学角度来看，这种通往最终产品的方法是有价值的，例如，建立虚拟目录。

2. A virtual shoe test bed considering different levels of user-footwear functional interaction (fitting, plantar pressures, shock absorption, flexibility, friction, and thermoregulatory aspects) and perceptual interaction (comfort) was developed.

开发了一个考虑不同级别用户–鞋类功能交互（合脚性、足底压力、减震、柔韧性、摩擦和热调节方面）和感知交互（舒适度）的虚拟鞋子测试平台。

3. Finally, the result of the virtual analysis for the complete footwear functional performance is linked to users' comfort perception through a regression model.

最后，通过回归模型将鞋类完整功能的虚拟分析结果与用户的舒适度感知联系起来。

4. In that sense, the designer receives worthy infromation on the footwear design at two levels: first, a functional performance characterization of different aspects of the user-footwear biomechanical interaction (fitting, pressure distribution, shock absorption) and, second, a prediction of the users' perception comfort when wearing such products.

从这个意义上说，设计师可以从两个层面获得对鞋类设计有价值的信息：

第一，用户与鞋类生物力学相互作用的不同方面的功能性能表征（合脚性、压力分布、减震），第二，预测用户穿着此类产品时的感知舒适度。

词 汇

computer-assisted engineering（CAE）：计算机辅助工程

finite elements analysis：有限元分析

thermoregulatory：体温调节的

CHAPTER 2

R&D of Functional Shoes

Lesson 1 Foot Pressure Measurements

For almost 150 years, the foot-loading behavior of the human foot has already been a subject of interest for **anatomists**, **orthopedists**, and biomechanists.[1]

Originally, the research aimed to understand the static and dynamic function of the human foot during standing, walking, running, and other athletic activities. Furthermore, gender differences in foot function, foot development during childhood, and foot problems in different patient populations were studied. The use of pressure distribution for the recognition and treatment of diabetic foot problems, and the effect of body weight and obesity on foot function and pain were studied by many research groups.

This basic research was performed by standing on and walking barefoot across **pressure distribution platforms**. In-shoe plantar pressure measurements are necessary to evaluate the effect of shoe construction on foot loading. **Resistive**, **capacitive**, and **piezoelectric** sensor technologies are commonly used to measure **pressure distributions**.[2] The availability of inexpensive **transducers**, rapid data acquisition, and efficient data processing has resulted in easy-to-use, commercially available pressure distribution instrumentation.[3] Therefore, pressure distribution measurements have become a standard clinical procedure for the diagnosis of foot problems and **therapeutic interventions**.

The most commonly used pressure plates have a size of about 50 cm in length and 40 cm in width just like the regular force plate sizes. However, pressure plates can differ in size and can even be smaller than 50 cm or be as big as 2 m nowadays.

The advantage of larger plates is that they can measure two consecutive steps. With two consecutive steps, additional information such as **step length**, **step frequency**, and **walking velocity** can be assessed. Another advantage is that subjects will be less focused on aiming to place their foot on the plate, which is often the case in smaller pressure plates.[4]

The **mid gait** protocol in which plantar pressure is measured during steady-state walking is the most favorable. Because **steady-state walking** requires a relatively long walkway and is time-consuming, the three-step protocol in which the third step is on the pressure plate is often used as a good alternative.[5] During the first two steps, subjects are mainly accelerating, which will have an effect on the plantar pressure pattern. The third step is usually a good indicator of normal walking. The three-step protocol appeared to have a high **reproducibility** not only during several measurements at one time but also between different days. It can be avoided that the subject is focusing on placing the foot on the plate by starting from a marked position, which automatically leads to the foot on the plate with a regular **walking pattern**.[6]

However, subjects with diseases that affect gait seldom have a regular walking pattern and the three-step protocol can then be replaced by the two-step protocol. The walking speed of patients is often reduced and the acceleration of the second step will be less compared to healthy subjects, and therefore it can be an appropriate alternative. Furthermore, the two-step protocol reduces the number of steps, which is beneficial for patients who often have difficulty walking and are quickly exhausted or who suffer from pain during walking.

Many studies have been performed to compare different step protocols or comparing one-step or two-step protocols with a mid gait method.

In general, no significant differences between protocols in **regional peak pressure** and/or **pressure-time integrals** were found. In diabetic patients, the two-step protocol required the least amount of repeated trials for obtaining reliable pressure and may be recommended for assessment of these patients. Hence, the three-step protocol might be the most favorable protocol, even though the one-step protocol has to be chosen due to the research group of interest.

To ensure that subjects walk normally over the plate during the measurement, it is recommended to watch them walk freely, since many subjects will alter their walking pattern when entering a laboratory setting.

重点及难点句

1. For almost 150 years, the foot-loading behavior of the human foot has already been a subject of interest for anatomists, orthopedists, and biomechanists.

近 150 年来，人类足部的负重行为已经成为解剖学家、骨科医生和生物力学家感兴趣的课题。

2. Resistive, capacitive, and piezoelectric sensor technologies are commonly used to measure pressure distributions.

电阻式、电容式和压电式传感器技术通常用于测量压力分布。

3. The availability of inexpensive transducers, rapid data acquisition, and efficient data processing has resulted in easy-to-use, commercially available pressure distribution instrumentation.

便宜的传感器、快速的数据采集和高效的数据处理令易于使用的商用压力分布仪器得以应用。

4. Another advantage is that subjects will be less focused on aiming to place their foot on the plate, which is often the case in smaller pressure plates.

另一个优点是受试者将不那么专注于将脚踩在压力板上，这通常发生在压力板较小的情况下。

5. Because steady-state walking requires a relatively long walkway and is time-consuming, the three-step protocol in which the third step is on the pressure plate is often used as a good alternative.

因为稳态行走需要相对较长的步道，而且很耗时，所以通常采用三步测试方案，其中第三步踏在压力板上，这是一个很好的替代方案。

6. It can be avoided that the subject is focusing on placing the foot on the plate by starting from a marked position, which automatically leads to the foot on the plate with a regular walking pattern.

通过从一个有标记的位置开始，可以避免受试者专注于将脚踩在板上，无意识地导致脚以有规律的行走模式踏在压力板上。

词 汇

anatomist：解剖员，解剖学家

orthopedist：骨科医生，矫形外科医生

pressure distribution platform：压力分布测试台

resistive：电阻式　　　　　　　　capacitive：电容式

piezoelectric：压电式　　　　　　pressure distribution：压力分布

transducer：传感器，感应器

therapeutic intervention：治疗性干预措施

step length：步长　　　　　　　　step frequency：步频

walking velocity：步速　　　　　　mid gait：中间步态

steady-state walking：稳定状态行走　reproducibility：可重复性

walking pattern：行走模式，步态模式

regional peak pressure：区域峰值压强

pressure-time integral：压强 – 时间积分

Lesson 2   Foot Orthoses

Any material placed between the sole of the foot and the inside of the shoe could be considered a **foot orthosis** since it will influence the forces acting on the foot;[1] however, a great deal of time has been dedicated to the classification and comparison of different types of foot orthosis. Orthoses may be classified by intended function, so an orthosis with an arch support and made of rigid materials will be expected to reduce **foot pronation** (a functional foot orthosis),[2] whereas a flat insole made from layers of **cushioning material** is intended to reduce forefoot pressures (an accommodative orthosis). As a further complication, all of these may or may not have additional wedges or raises under the heel and forefoot to influence foot motion and load distribution,[3] and the features of a functional foot orthosis and an accommodative foot orthosis are commonly combined.

"Which orthosis type is the best?" is a common question but there is no simple answer; indeed, it is the wrong question to ask. The pertinent question is "Which shape and material of foot orthosis works are the best for this specific patient?". In most cases, a **prefabricated orthosis** can be found that offers appropriate clinical outcomes.

It is critical when evaluating evidence relating to the efficacy of foot orthoses that conclusions are not incorrectly extrapolated to all orthoses of a similar design.[4] Evidence that one specific casted orthosis is better at controlling pronation of the foot is not evidence that all casted orthoses perform this function better than prefabricated orthoses.[5] The data relate only to the specific orthosis tested. A different **casted orthosis** could have the opposite result when compared with an alternative prefabricated orthosis. So, do not attempt to classify your practice by the manufacturing method of "type" of orthosis, but rather consider which orthosis offers what your patient requires.

Foot orthoses are used for a wide range of clinical symptoms but underlying these symptoms are three biomechanical objectives:

1. To alter foot motion.

2. To alter stress experienced by internal hard and soft tissues.

3. To alter the distribution and magnitude of load applied to the plantar surface.

Foot orthoses have the potential to reduce symptoms associated with many localized mechanical problems and those associated with systemic diseases. Further to this, they can improve function, mobility and a person's ability to carry out all the activities of daily living. For foot orthoses to be effective in achieving these benefits, clinicians need to be able to choose the right design and materials. As there is an **extricable** relationship between the foot, the orthoses and the footwear that contains the two, footwear design and usage have a direct impact on the potential for the foot orthoses to be effective.[6] The following chapters will provide the clinician with the knowledge and understanding that is required if we are to achieve the right combination of orthoses and footwear. Further to this, an understanding of the role of footwear as being more than protection is required if we, as clinicians, are to understand the impact of what we consider an intervention and, from the patient's perspective, an item of clothing that is visible and can define who they are.[7] This view of footwear has become established over centuries; therefore, the next chapter will give a brief outline of the evolution of footwear design and purpose.

重点及难点句

1. Any material placed between the sole of the foot and the inside of the shoe could be considered a foot orthosis since it will influence the forces acting on the foot.

位于足鞋之间的任何材料都可以被看作是足部矫形器，因为它会影响作用在脚上的力。

2. Orthoses may be classified by intended function, so an orthosis with an arch support and made of rigid materials will be expected to reduce foot pronation (a functional foot orthosis).

矫形器可以根据预期功能进行分类，因此由硬质材料制成的具有足弓支撑的矫形器有望减少足内翻（即功能性足部矫形器）。

3. As a further complication, all of these may or may not have additional wedges or raises under the heel and forefoot to influence foot motion and load distribution.

更复杂的是，所有这些矫形器在后跟和前掌部位可能有也可能没有附加的楔形或凸起，从而影响足部运动和压力分布。

4. It is critical when evaluating evidence relating to the efficacy of foot orthoses that conclusions are not incorrectly extrapolated to all orthoses of a similar design.

重要的是，当评估与足矫形器疗效相关的证据时，不能错误地推断出结论适用于所有类似设计的矫形器。

5. Evidence that one specific casted orthosis is better at controlling pronation of the foot is not evidence that all casted orthoses perform this function better than prefabricated orthoses.

有证据表明某一种铸造的矫形器能更好地控制足旋前，但并不代表所有的铸造矫形器都比预制矫形器能更好地控制足旋前。

6. As there is an extricable relationship between the foot, the orthoses and the footwear that contains the two, footwear design and usage have a direct impact on the potential for the foot orthoses to be effective.

由于脚、矫形器和包含两者的鞋之间均可相互分离，鞋类的设计和使用直接影响着足部矫形器是否有效。

7. Further to this, an understanding of the role of footwear as being more than protection is required if we, as clinicians, are to understand the impact of what we consider an intervention and, from the patient's perspective, an item of clothing that is visible and can define who they are.

此外，作为临床医生，如果我们想要了解我们所认为的干预措施的影响，以及从患者的角度来看，鞋子是一件可见的、可以确定他们身份的服饰，就需要了解鞋类的作用，而不仅仅是起保护作用。

词　汇

foot orthosis：足部矫形器　　　　　foot pronation：足旋前，足内翻

cushioning material：弹性垫料　　　prefabricated orthosis：预制矫形器

casted orthosis：（浇注）成型矫形器　extricable：可解脱的，可分离的

Lesson 3   Function Foot Orthoses

Functional shoes face different audiences and different wearing scenarios, so the development of functional shoes has its characteristics. The research and development of functional shoes can start from multiple levels such as material modification and technical assistance, etc. According to the needs of different target groups, certain technical means to make shoes in a certain aspect with outstanding functionality which is the purpose of functional shoe research and development.

Material modification

The development of occupationally functional footwear with anti-microbial outsole materials has been proposed by incorporating footprint-side removable inserts (manufactured by 3D printing) with the addition of silver nanoparticles (AgNPs) to provide aseptic properties to this type of shoe from its manufacture for application in the food manufacturing industry.

Several studies have prepared **SEBS**/EVA foamed shoe materials by using environmentally friendly inert gases in the super/subcritical state to provide low shrinkage, plasticity, high elasticity, and ease of processing.[1]

Application of **phase change materials (PCMs)** to footwear improves the thermal comfort of human feet exposed to cold and wet weather. The warming effect provided by PCMs depends primarily on the amount of PCM applied, the temperature gradient between the curing temperature of the PCM and the ambient temperature, and the area covered.[2] Further research is still needed to explore the possibility of incorporating PCM into warm footwear and military footwear.

Technical assistance

Relying on **flexible electronic sensors**, wearable plantar pressure monitoring provides a viable solution for applications such as foot disease diagnosis and pressure monitoring. Plantar pressure monitoring is often available in the form of both pressure-detecting socks and monitoring shoes. The difference between the two is that the former uses a one-piece forming process when knitting, interweaving conductive yarns with ordinary yarns; the latter side integrates flexible pressure

190

sensor components directly into the insole, by connecting the signal acquisition with transmission and processing modules by means of thin wires to realize data acquisition and processing.[3]

Accelerometer and gyroscope-based navigation paths have been widely used in many fields such as motion evaluation and location tracking. By processing and analyzing the data acquired by the **accelerometer** and **gyroscope**, the wearer can be located, and when a person falls or other unexpected situations are encountered, the address of the person can be quickly informed to his relatives and friends to facilitate the removal of danger.

Harvesting energy from shoes is receiving increasing attention from researchers. With the miniaturization of electronics, wearable technology has enabled shoes with more innovative features. The rationale for using shoes to harvest energy lies in the fact that energy can be harvested through vibrations, compression or bending generated in footwear when the wearer performs athletic activities, such as walking.[4] **Piezoelectric lead zirconate titanate (PZT)** which is a ceramic material, and **polyvinylidene fluoride (PVDF)** which is a plastic material are commonly used as energy harvesting elements.[5]

In general, the development of functional shoes cannot be separated from the innovation of materials and technology. In order to develop functional shoes, we need to fully understand the needs of target users and reasonably analyze their real application scenarios. Material modifications, structural improvements and process enhancements can be considered to empower shoe and boot products and provide greater convenience and comfort for the wearer.

重点及难点句

1. Several studies have prepared SEBS/EVA foamed shoe materials by using environmentally friendly inert gases in the super/subcritical state to provide low shrinkage, plasticity, high elasticity, and ease of processing.

一些研究利用环境友好的惰性气体在超 / 亚临界状态下制备丁苯橡胶 / 乙烯 – 醋酸乙烯共聚物发泡鞋材，该材料具有低收缩、可塑性、高弹性、易加工等优点。

2. The warming effect provided by PCMs depends primarily on the amount of PCM applied, the temperature gradient between the curing temperature of the PCM and the ambient temperature, and the area covered.

PCM（相变材料）提供的升温效应主要取决于相变材料的应用量、相变材料的固化温度与环境温度之间的温度梯度，以及所覆盖的面积。

3. The difference between the two is that the former uses a one-piece forming process when knitting, interweaving conductive yarns with ordinary yarns; the latter side integrates flexible pressure sensor components directly into the insole, by connecting the signal acquisition with transmission and processing modules by means of thin wires to realize data acquisition and processing.

两者的区别在于前者在编织时采用一体式成型工艺，将导电纱与普通纱交织；后者将柔性压力传感器组件直接集成到鞋垫中，通过细线将信号采集与传输处理模块连接起来，实现数据采集与处理。

4. The rationale for using shoes to harvest energy lies in the fact that energy can be harvested through vibrations, compression or bending generated in footwear when the wearer performs athletic activities, such as walking.

使用鞋子收集能量的基本原理在于，当穿着者进行运动（如走路）时，可以通过鞋类产生的振动、压缩或弯曲来收集能量。

5. Piezoelectric lead zirconate titanate (PZT) which is a ceramic material, and polyvinylidene fluoride (PVDF) which is a plastic material are commonly used as energy harvesting elements.

压电锆钛酸铅（PZT）是一种陶瓷材料，聚偏氟乙烯（PVDF）是一种塑料材料，通常用作能量收集元件。

词 汇

SEBS：是 styrene-ethylene/butylene-styrene 的缩写，指聚苯乙烯－聚（乙烯／丁烯）－聚苯乙烯嵌段共聚物

phase change materials (PCMs)：相变材料

flexible electronic sensors：柔性电子传感器

accelerometer：加速度计

gyroscope：陀螺仪

harvesting energy：能量采集

piezoelectric lead zirconate titanate (PZT)：压电锆钛酸铅（PZT）

polyvinylidene fluoride (PVDF)：聚偏氟乙烯（PVDF）

CHAPTER 3

Technology Trends

Lesson 1   Footwear, Balance, and Falls in the Elderly

As the world's population is aging, falls among the elderly constitute a serious public health problem. It is estimated that one in three people living in the community aged 65 or over, experiences at least one fall per year.

There are many identified risk factors for falls, including intrinsic risk factors associated with age-related degeneration of the balance and **neuromuscular systems** and medical conditions and extrinsic risk factors including environmental factors.

Most falls occur during motor tasks, and footwear has been identified as an environmental risk factor for both indoor and outdoor falls. By altering **somatosensory feedback** to the foot and ankle and modifying **frictional conditions** at the shoe-sole/floor interface, footwear influences postural stability and the subsequent risk of **slips, trips**, and falls.[1]

Although shoes are primarily devised to protect the foot and facilitate propulsion, the influence of fashion on footwear design throughout the ages has compromised the natural functioning of the foot. As a result, little is known about what constitutes safe footwear for the elderly when undertaking activities in and around the home. Because footwear appears to be an easily modifiable fall risk factor, it is imperative to identify the specific shoe features that might facilitate or impair balance in the elderly so as to design targeted fall prevention interventions and provide evidence-based recommendations.[2]

Optimal footwear for the elderly

The question "What is the safest footwear for the elderly who have fallen or are at risk of falling?" raised by the **American and the British Geriatrics Societies** and the **American Academy of Orthopedic Surgeons** does not have a definitive answer, despite substantial advances in the field of footwear and falls research.

There is now sufficient **epidemiological** evidence to suggest that the elderly should wear appropriately fitted shoes both inside and outside the house, since walking barefoot and in socks indoors are the footwear conditions associated with the greatest risk of falling.

Given the high prevalence of foot problems among the elderly,[3] there is a need for shoes that can accommodate the extra **girth of the forefoot** for individuals with hallux valgus, increased depth of the toe box for those with lesser **toe deformities**, and footwear with compliant material able to stretch around swollen feet. Nevertheless, providing that they have adequate fastenings, shoes with a low square heel, a sole of medium hardness (**shore A**-40), and a high collar provide optimal balance in healthy older community-dwelling people during functional balance and stepping tests and walking and stopping on a range of surfaces.[4] This shoe type is therefore recommended as safe for wearing in and around the home, on dry and wet **linoleum** floors, as well as on irregular terrain.

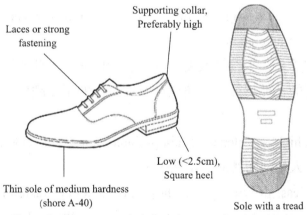

Laces or strong
fastening

Supporting collar,
Preferably high

Thin sole of medium hardness
(shore A-40)

Low (<2.5cm),
Square heel

Sole with a tread

Figure 1   Recommended shoe features for the elderly

In contrast, shoes with an elevated heel are deemed unsafe because of their detrimental effects on balance and gait patterns and because of their reported lack of comfort and stability. The elderly should also be advised not to wear shoes with soft soles (sole hardness less than shore A-33), as these can alter balance control during challenging gait tasks.[5] A tread sole may further prevent slips on wet and slippery surfaces.

Prevention of falls should also include education of the elderly and their carers/family (for those housebound or institutionalized) regarding these footwear recommendations as it is likely that financial and comfort aspects currently outweigh safety considerations when the elderly purchase shoes.[6] Future directions for research should include clinical studies assessing **slip-resistant features** of the sole that can prevent indoor slipping. Finally, the potential benefits of somatosensory stimulating insoles on postural control should be explored further.

重点及难点句

1. By altering somatosensory feedback to the foot and ankle and modifying frictional conditions at the shoe-sole/floor interface, footwear influences postural stability and the subsequent risk of slips, trips, and falls.

通过改变对足部和踝关节的体感反馈以及改变鞋底/地板界面的摩擦条件，鞋类可影响姿势的稳定性以及随后的滑倒、绊倒和跌倒的风险。

2. Because footwear appears to be an easily modifiable falls risk factor, it is imperative to identify the specific shoe features that might facilitate or impair balance in the elderly so as to design targeted fall prevention interventions and provide evidence-based recommendations.

由于鞋类似乎是一个容易改变的跌倒风险因素，因此必须确定可能促进或损害老年人平衡的鞋类特征，以便设计有针对性的预防跌倒干预措施，并提供有依据的建议。

3. Given the high prevalence of foot problems among the elderly, ...

考虑到老年人足部问题的高发性……

4. Nevertheless, providing that they have adequate fastenings, shoes with a low square heel, a sole of medium hardness (shore A-40), and a high collar provide optimal balance in healthy older community-dwelling people during functional balance and stepping tests and walking and stopping on a range of surfaces.

然而，在社区居住的健康老年人进行平衡功能和步态测试时以及在各种表面上行走和停止时，足够的束紧方式，低方跟、中等硬度鞋底（硬度为 40 度）

和高帮的鞋子可以提供最佳的平衡。

5. ... , as these can alter balance control during challenging gait tasks.

……，因为这可能会改变（老年人）在挑战性步态任务中的平衡控制。

6. ... as it is likely that financial and comfort aspects currently outweigh safety considerations when the elderly purchase shoes.

……，因为目前老年人在购买鞋类时，经济和舒适方面的考虑可能超过对安全的考虑。

词 汇

neuromuscular system：神经肌肉系统

somatosensory feedback：体感反馈

frictional condition：摩擦力状况，摩擦力条件

slip：滑倒

trip：绊倒

American and the British Geriatrics Societies：美国与英国老年病学会

American Academy of Orthopedic Surgeons：美国骨科医学会

epidemiological：流行病学的

girth of the forefoot：前足围度

toe deformity：脚趾畸形

shore A：邵尔 A（硬度计量）

linoleum：油毯，油地毡

slip-resistant feature：防滑性能

Lesson 2   Design & Production of Sustainable Shoes

Designing and producing sustainable shoes can be difficult but is an important factor to consider in today's world of increasing climate change and climate change activism. We must think ahead, and start with our designs and material selection, to reduce our carbon footprint as consumers and manufacturers in the shoemaking industry. The modern mass-produced shoe, made of textiles, leather, plastic, and rubber parts all glued and sewn together is not environmentally sustainable.

1. The global footprint of shoemaking

The modern shoe is very difficult to recycle. The shoe factory workers secure the upper parts with stitching and firmly bond the outsole parts with PU cement. Used shoes and sneakers are almost impossible to break down into useful components for recycling. The manufacturing of these components themselves consumes vast amounts of water and energy while creating mountains of post-industrial and post-consumer waste.

2. Your choices for sustainable shoe production

Footwear designers, shoe developers, product managers, and factories can make choices to help reduce the negative environmental and social impacts of shoe production. There are no magic shoe materials or production techniques that can make a shoe entirely green, sustainable, or ethical. Depending on your own environmental and social priorities, there are many options available.

We will consider different aspects of shoe production that can make your shoes more or less environmentally sustainable:

Sustainable shoe material selection

Environmentally friendly footwear production processes

Waste reduction in footwear manufacturing

3. Organic vs. man-made textiles

If your priority is drinking water preservation, then using man-made textiles is a better choice over cotton and other natural textiles. Both the cotton and man-made fibers require large quantities of water for dyeing processes. Fortunately, the water in an industrial facility can be recovered, recycled, and reused in a closed-loop system. In China, some local governments have forced textile dyeing houses to relocate into industrial estates with controlled water purification facilities.

The process of growing cotton, especially organic cotton, consumes enormous quantities of water that is not reused in a **closed-loop**. Some studies estimate that more than 700 gallons (2,700 liters) of water is required to make the cotton for one cotton T-shirt! Yes, once we use water in the cotton fields, it does return to nature, but it is no longer available to drink or grow food crops.

On the flip side of cotton and natural fibers, is the production of man-made polymer-based fabrics such as nylon or polyester.[1] The amount of water required to make these fibers is radically less, but the energy requirement is higher, and there is a greater danger of water contamination from petrochemicals.

4. Natural vs. man-made "leather"

Natural leather from animal hides also requires large amounts of water. According to studies done by a major leather producer, raising animals and processing their hides requires over 264 gallons (1,000 liters) of water to produce two square feet of leather. Two square feet of leather is enough to make just one pair of shoes. Raising animals and then processing leather into hides has a two-fold effect on the environment: the contaminated and harmful agricultural run-off plus water which contains hazardous tanning byproducts. Water aside, the treatment of animals is a significant concern for vegan customers while the production of human-made imitation leather is not regularly considered a moral hazard.[2]

As with fabrics, the human-made alternatives to natural leather have their environmental costs. Human-made synthetic materials are very often layers of polyester fabric, foams, and fibers that are fused together.[3] These layers are nearly impossible to separate once the shoe has reached the end of its lifespan. Producing synthetic materials also consumes energy, and the danger of water contamination from petrochemicals is high.

5. Natural vs. synthetic rubber

Again, the choice of rubber compounds comes down to a choice of your environmental priorities. Natural rubber production leads to increased deforestation in Southeast Asia and reduces the amount of land available for food cultivation. Synthetic rubber production requires a combination of **styrene** and **butadiene**. Both are petrochemicals refined from crude oil. The production of these compounds requires significant energy inputs, and both are byproducts of oil production.

Although there are many material options, each comes with either an environmental or a social cost. When you make sustainable shoes, you need to decide where your priorities lie.

6. Shoe materials with recycled content

Another way to reduce the overall environmental impact of footwear production is to specify some of the many footwear materials made with recycled content.[4] When reviewing materials for sustainable shoes, it is important to understand the difference between post-consumer and post-industrial waste. Many industrial processes create waste or scrap inside the factories. However, the supply factory will recover and reprocess these materials into the finished materials. For instance, the injection molding supplier will regrind, and re-mold wasted materials.[5] Fabric factories will chop and re-purpose textile fibers. For many factories, this is a simple and smart way to save money. Factories will collect other post-industrial waste and send it out for reprocessing into various other products.

Post-consumer recycled content is produced when the raw materials are recovered from the waste stream after use.[6] These products may cost more, as the materials may require complicated sorting, cleaning, and reprocessing. The amount of post-consumer content in a product depends on the physical properties required. Usually, higher physical test standards will demand lower post-consumer content. Fabrics are now available with 10% to 70% recycled content.

Textile factories now make many woven and knit fabrics made with post-consumer recycled **PET** plastic fibers. Shoe-lasting board suppliers now produce Strobel materials with both post-industrial and post-consumer waste. Paper fiber-based lasting boards often contain over 50% post-consumer waste.

Additionally, foam factories can now supply shoe footbeds and linings made with post-consumer and post-industrial recycled foam materials and new **biodegradable additives** are available that allow plastic to degrade in decades rather than centuries.[7]

重点及难点句

1. On the flip side of cotton and natural fibers, is the production of man-made polymer-based fabrics such as nylon or polyester.

与棉花和天然纤维相对的另一方面，是人造聚合物基织物如尼龙或聚酯的生产。

2. Water aside, the treatment of animals is a significant concern for vegan customers while the production of human-made imitation leather is not regularly considered a moral hazard.

除水之外，如何对待动物是素食者关注的一个重要问题，而人造皮革的生产通常不被认为有道德风险。

3. Human-made synthetic materials are very often layers of polyester fabric, foams, and fibers that are fused together.

人造合成材料通常由多层涤纶织物、发泡材料及纤维融合在一起。

4. Another way to reduce the overall environmental impact of footwear production is to specify some of the many footwear materials made with recycled content.

另一种减少鞋类生产对环境影响的方法是在许多鞋类材料中指定使用某些可回收利用的材料。

5. For instance, the injection molding supplier will regrind, and re-mold wasted materials.

例如，注射成型供应商将废料重新研磨并重新模压成型。

6. Post-consumer recycled content is produced when the raw materials are recovered from the waste stream after use.

消费后回收物是指在产品使用后从废物流中回收原料时产生的。

7. Additionally, foam factories can now supply shoe footbeds and linings made with post-consumer and post-industrial recycled foam materials and new biodegradable additives are available that allow plastic to degrade in decades rather than centuries.

此外，泡沫工厂现在可以提供由消费后和工业生产后回收的发泡材料制成的鞋垫和衬里，以及新的生物可降解添加剂，使塑料可以在几十年而不是几个世纪内降解。

词 汇

closed-loop：闭环系统，闭合回路
styrene：苯乙烯
butadiene：丁二烯
PET：聚对苯二甲酸乙二醇酯
biodegradable additive：可生物降解的添加剂

Glossary

a new line of shoes：新款鞋系列

abnormality：异常，畸形，变态

abrasion mark：擦痕，磨痕

academic debates：学术辩论会

accelerometer：加速度计

achilles tendonitis：跟腱炎

acted as：充当……

action leather：二层革，合成革

action or coated leather：二层革，移膜革

adhesive：胶黏剂

age：老化

aglet：鞋带封头

air-cushioned insole：气垫鞋垫

American Academy of Orthopedic Surgeons：美国骨科医学会

American and the British Geriatrics Societies：美国与英国老年病学会

anatomical：解剖的

anatomist：解剖员，解剖学家

aniline：苯胺

anterior：前部的；前面的

aporn：前帮盖

appendage：下肢

arch：足弓

archiving and retrieval：存档和检索

arthritis：关节炎

articulation：关节

artisan：手艺人

assembly line：装配线

attire：服装，盛装

back part forming machine：后帮 / 主跟（预）成型机

back：脚背，跗背，背衬

backbone：背脊线

backing：衬里，背衬

backpart stability：（鞋）后跟部位稳定性

ball：拇趾球，跖趾部位

ballet flats：软帮平底鞋

ball-of-the-foot：脚掌

ballroom shoes：舞鞋

barbed wire：铁丝网

basket：网眼

bath clog：木质浴室鞋，木底鞋，木屐

be well placed to：处于有利地位

beam cutting press：横梁式裁断机

bed：使平铺于……上，使紧伏于

belly：边腹部

bent：弯曲

bias：倾斜，与……成一定的角度

binding：沿口，滚边，包边材料

biodegradable additive：可生物降解的添加剂

biomechanical：生物力学的

Blake (McKay welt) construction：缝内线结构

Blake construction：缝内线结构

Blake stitch：（缝）内线

Blake/rapid construction：内外线透缝结构

blemish：瑕疵

block copolymer：嵌段共聚物

Blüchers (Derbys)：外耳式鞋，德比鞋

board lasted (stuck on，cement-lasted) construction/ Compo Process：绷楦结构，绷楦法

Bologna construction：博洛尼亚结构

bond welt：假沿条（鞋）

bond：结（胶，粘）合

bottom：合底，制底

brass：黄铜

break：分割，断裂

breathability：透气性

breathable：透气的

bristol paper：高级绘图纸

brogues：布洛克鞋

brushing：涂刷，刷浆

buckle：鞋钎，鞋扣

buff：砂磨

bump toe：高头（楦）

bunion：拇囊炎

bunionette：小趾囊炎

butadiene：丁二烯

button：鞋扣

CAD system：计算机辅助设计系统

calcaneus：跟骨

calf muscle：小腿肌肉

calfskin：小牛皮

callus：皮肤的硬结，老茧，胼胝

canvas：帆布

capacitive：电容式

cap-toes：包头鞋

care instruction：保养说明，使用说明

cast aluminum：铸铝

casted orthosis：（浇注）成形矫形器

casual shoes：休闲鞋

cementing：刷胶

certified：达到标准的，有合格证书的

chakris：象牙质夹趾拖鞋

charring：碳化

city rubber sole：城市鞋底，仿皮底

claw toe：爪状趾

cleated shoe：钉鞋，防滑鞋

clicker：裁断工，裁断机

clicking or cutting department：裁断车间

closed-loop：闭环系统，闭合回路

closing or machining department：机缝车间

closing：缝帮

closure：闭合件，闭合方式

CNC（Computer Numerical Control）：电脑数控

coalescing property：黏合性能

coating：涂层，涂饰

collar foam：鞋口泡棉，鞋领口泡棉

collar：（运动鞋等的海绵）护口，鞋口，装饰性沿口

colorfastness：色牢度

commando sole：登山鞋底

complication：并发症

compression molded EVA：模压乙烯 – 酸酸乙烯酯共聚物（EVA）

compression molding：模压

compression set：压缩变形

computer-assisted engineering（CAE）：计算机辅助工程

conductive：导电的

configurator：配置，配置程序

confirmation sample：确认样（品）

constriction：束缚

conventionally：传统地

cook：热硫化

corn：鸡眼

cornerstones：基石

corrected and pigmented grain：修面染色革

cosmetic defect：外观缺陷

court shoes：船鞋

crazy horse：疯马皮

crepe sole：绉胶底，生胶底

critical defect：严重缺陷

crooked stitch：缝线不直

cross section：横截面

crosslink：交联

cuboid：骰骨

cuneiform：楔状骨

cure：硫化，熟化

curing time：硫化时间，（交联）反应时间

curve-lasted shoe：弧线鞋

cushioning material：弹性垫料

cushioning wedge：缓冲内插（楔状物）

cushioning：缓冲

customization：定制，专用化

customized：定制的

cutter or clicker of leather：皮革裁断工

cutting board：裁断垫板

cutting chamber：切割台

cutting die：裁断刀模

cutting knives：（手工）裁刀

cutting loss：裁断损耗

cutting operation：裁断操作

cutting press/clicker cutting machine：裁断机

daint：小巧漂亮的

decorated：装饰的，修饰的

decoration perforation：装饰花孔（眼）

decoration seam：装饰缝

defect：缺陷

degumming：脱胶

denier：旦尼尔（单位）

design orientation and position：图样方向和位置

die-cut：刀模冲切

digitization：数字化

directional term：方向术语

distal：末梢的，末端的

doddy：短绒

double knit：双面针织物

drawing：标记

durability：耐久性，耐用性，耐用度

edge beating：锤敲边口

edge folding：折边

edge of the skin：皮革边缘（部位）

edge treatment：（皮料）边口处理

edging：熨烫边口

elaborately：精巧地

elastic-side shoes：松紧带鞋

elite：上层人士，尖子，精英

emboss：压花，装饰

embossing：压印

environmentally sound：环境友好的，环保的

epidemiological：流行病学的

EVA (ethyl vinyl acetate)：EVA（乙烯－醋酸乙酯）

excavation：挖掘

excessive roll inward：过度内旋

exterior：外观，外部，外貌

extricable：可解脱的、可分离的

eyelet：鞋眼

fabrication technique：制造技术，制作工艺

facility：设备，工具，设施

fastening：紧固件、束紧装置

fat pad：脂肪垫

fat wrinkle：肥皱

feather edge：楦底棱

feed abrasion mark：进料磨损痕迹

fibula：腓骨

fibularis brevis muscle：腓骨短肌

finishing department and shoe room：成品车间

finishing：整理，涂饰

finite elements analysis：有限元分析

fit：合脚

fitting check：合脚性检验

fitting fastening：装配紧固件

flare：喇叭形

flat foot：扁平足

flat machine：平板缝纫机

flexibility：弹性，柔性

flexibility：屈挠性，柔韧性，柔软度

flexible electronic sensors：柔性电子传感器

FlyKnit：飞织

foam insole：发泡（材质）鞋垫

foam：海绵，发泡材料，泡沫

foot orthosis：足部矫形器

foot pronation：足旋前，足内翻

foot tracing：足印

foot-activated press：脚踏式压机（画线机）

football cleat：足球（钉）鞋

footbed：鞋垫

footprint：脚印

force lasting construction (Strobel-stitched method，sew-in sock)：闯楦工艺

forefoot：前足

forepart lasting machine：绷尖机

forepart：前段，前帮

foxing tape：外胶条，围条

fragment：碎片

freehand drawing：徒手画，示意图

friction：摩擦，摩擦力

frictional condition：摩擦力状况，摩擦力条件

full grain aniline：全粒面苯胺革

full grain pigmented (top grain)：全粒面染色革（头层革）

full skin：整张革

fuse：熔合

gait cycle：步态周期

gel sheet：凝胶垫

generalization：概括

geometric modeling tools：几何建模工具

getas：日式木屐

gimping：切锯齿形边口

girth：围度

girth of the forefoot：前足围度

glue：用胶水将物体黏合，胶水

glycerin：甘油

golden sample：标（准）样（品）

Goodyear welt construction：固特异缝沿条结构

grain side：粒面

grip：抓地力

groove：容线槽

gummy：有黏性的

gyroscope：陀螺仪

hallux valgus：拇趾外翻，拇外翻

hammer toe：锤状趾

hand cutting：手工裁断

hand span：用手丈量，掌距

hand stitching：手工缝

harvesting energy：能量采集

heart of the skin：皮心（部位）

heat reflecting coatings：热反射涂层

heat setting：加热定型

heel：脚后跟，鞋后跟

heel attachment：装跟牢度，鞋跟附着强度

heel cap/heel tip：鞋跟面皮

heel cradle：后跟支架

heel cup：后跟垫，跟杯，跟部凹形插件

heel notch：后帮上口（V字型口）

heel spur：足跟刺痛

heel strength：鞋跟强度，

hem：下摆，折边，底边

heterogeneous material：非均匀材料

hide and skin：皮革（hide一般指面积较大的，如牛皮；skin一般指面积较小的，如羊皮）

high arch foot：高弓足

high-arch / pes cavus：高足弓

high-end：高端的

hinge last：弹簧铰链楦

hit the ground：落地（指概念被人们知晓）

homogeneous material：均质材料

hot melt cement：热熔胶

identification marking/stamping：点标志点

imitation moccasin：仿莫卡辛鞋

imperfection：缺陷

imprint：足印

incorrect sizing：错码

initial bond strength：初始黏合强度

injection molded：注塑的（外底）

inside side：里怀（部件）

insole board：内底板

insole：鞋垫

instep：脚背，跗面

insulate：绝缘，隔离，隔热

interlock：连锁

intractable plantar keratosis：顽固性足底角化症

irregularity：不规则性

jersey：平针织物，绒布

joint disorder：关节紊乱

kharrows：金属质夹趾拖鞋

kidskin：小山羊皮

kinesiologist：人体运动学家

labeling requirements for country of origin：原产国标签要求

laces：鞋带

laminated：层压，复合

laser cut：激光裁断

lasting and making department：绷帮和制底车间

lasting margin：绷帮余量

lasting：绷帮

lateral longitudinal arch（LMA）：外侧纵弓

lateral：外侧的

latex & cork：乳胶和软木

leak：渗入，透水

lift：后跟皮；插跟

light sole：轻质鞋底

line drawing：线稿

lining：衬里

linoleum：油毯，油地毡

locally：局部的

lock-stitch：锁缝

longitudinal arches：足纵弓

low-arch / pes planus：低足弓

lower side：（部件的）下口边

lower：底部件

Lycra：莱卡

major defect：主要缺陷

mallet toe：槌状趾

manual：手工的

mass customization：大规模定制

material exploitation / utilization：材料的开发 / 利用；排料方案

medial longitudinal arch（MLA）：内侧纵弓

medial：内侧的

melt：熔化，塑化

mesh：网状材料

metal shank：金属勾心

metatarsal bone：跖骨

metatarsal I–III：第 1 至 3 跖骨

metatarsal IV and V：第 4 和第 5 趾骨

metatarsus：跖骨

mid gait：中间步态

midsole filler：中底填充物

miles per gallon（MPGs）：每加仑燃料所行驶的英里数

miles per hour（mph）：每小时所行驶的英里数

minor defect：轻微缺陷

moccasin：莫卡辛软皮鞋（原为美洲土著所穿），皱头式软帮鞋，包子鞋

moc-toe：皱头，莫卡辛鞋头

model shoe-last：母楦

moisture wicking fabric：吸湿面料

molded construction：模压工艺

monk-straps：搭扣鞋，孟克鞋

MOQ (Minimum Order Quantity)：最小订货量

morphology：形态学

moss：苔藓

motion control：运动控制

mules：穆勒鞋，皮拖

natural crepe rubber：皱片胶（由天然橡胶加工而成）

navicular：舟状骨

neck：颈部

neoprene rubber：氯丁橡胶

nest：套排，嵌套

neuromuscular system：神经肌肉系统

normal-arch：正常足弓

Norwegian Welt / Storm Construction：压条结构

nubuck：正绒面革

nucleation：成核

nylon：尼龙

one-off shock loading：一次性冲击载荷

one-piece：一体式的

open stitch：浮线

orthopedic：矫形的，骨科的

orthopedist：骨科医生，矫形外科医生

orthotics：足部矫形器，矫形鞋垫

outside side：外怀（部件）

overlocking machine：锁缝机

overpronated：内翻的

over-pronation：过度内旋，外翻

over-supinated：外翻的

Oxfords：牛津鞋

padded Stroble sock：加衬锁缝鞋套（闯楦成型法）

padding：衬垫，填料

padukas：木质夹趾拖鞋

Paleolithic：旧石器时代

palm：棕榈树

Pantone：潘通，潘通色彩系统

parametric：参数的，参量的

patent leather：漆革，漆皮

pattern cutter：样板切割机

pattern lay out：排料（在材料上排布下料样板）

pattern lines：图案线条

patterned material：有图案、花纹的材料

patterns' lay out system：排料方案

peep toe high heels：鱼嘴高跟鞋，前空式高跟鞋

penetration：渗透性

perforating：冲孔，打孔

perforations：孔眼，花孔

performance footwear：功能鞋

permeable：可渗透的，可透过的

PET：聚对苯二甲酸乙二醇酯

phalange：趾骨，指骨

phalanx：趾骨，指骨

phase change materials (PCMs)：相变材料

photoelectric cell：光电电池

photorealism：真实感

piezoelectric lead zirconate titanate (PZT)：压电锆钛酸铅（PZT）

piezoelectric：压电式

pigskin：猪皮

pillars：支柱

pincer：（绷帮机上的）夹钳

plain material：单色的／无图案花纹的光面材料

plain：平纹

plain-toes：素头式

plantar calcaneonavicular ligament：足底跟舟韧带

plantar fasciitis：足底筋膜炎

plasticizer：增塑剂

pliable：柔韧的，柔软的

pocket：空腔，夹袋

podiatrist：足科医生

pointe shoes：芭蕾舞鞋

polychloroprene adhesive：氯丁橡胶胶黏剂，氯丁胶

polyester spandex：涤纶氨纶

polyester：聚酯纤维，涤纶

polyethylene：聚乙烯

polypropylene：聚丙烯，丙纶

polyurethane adhesive：聚氨酯胶黏剂

polyurethane：聚氨酯

polyvinylidene fluoride (PVDF)：聚偏氟乙烯（PVDF）

pore：小孔，毛孔

position：定位

post machine：高桩柱缝纫机

post：缝纫机柱，架

posterior：后部的；后面的

poured PU (polyurethane foam)：浇注 PU（聚氨酯发泡材料）

preassembling：镶接

prefabricated orthosis：预制矫形器

premature wear：过早磨损

pre-molding：预成型

pressure distribution platform：压力分布测试台

pressure distribution：压力分布

pressure-time integral：压强 – 时间积分

pretty much the same：相差无几

primer coat：底涂层

primer：处理剂

profit or loss：盈利或亏本

protruding nail：露钉，突出的钉子

proximal：近端的，近侧的

puckered seam：鼓包，缝合不平伏

pullover machine：套楦机

punching：冲孔

quarter：后帮

quill：大翎毛，羽茎，（豪猪、刺猬的）刚毛

radio frequency welding (RF welding)：射频焊接

raw edge：切割边，断口

rayon：人造丝，人造纤维

reassert：恢复原状

rebound：回弹

regional peak pressure：区域峰值压强

reinforcement：定型材料，补强衬料，增（补）强（加固）件

reproducibility：可重复性

reproduction：复制品，仿制品

resin rubber：树脂橡胶

resistive：电阻式

resole：更换外底

rib：内底埂条

ribbed：有棱纹的，有凸纹的，罗纹针织物

ribbon：丝带

ridgeway sole：山脊路鞋底

riding gear：马具

ripstop：格子

rite：典礼

ritual：仪式

rocker sole：摇杆底

rocking test：摇晃测试（带跟鞋的稳定性测试）

roll：辊压，辊筒

rosette：花结

roughing：砂磨

rubber cupsole：橡胶成型底

runner：鞋底边沿

run 24-7：全天候运行

rust-free metal：防锈金属

sabots：木屐，木鞋，木底皮鞋

safe in bonding：黏合牢固

sagebrush：灌木蒿

salability：畅销，出售

sample cutting：样品切割

sandal：凉鞋

satin：缎纹

saturated：饱和的

SBR (Styrene butadiene rubber)：丁苯橡胶

scoop shoe-last：活盖鞋楦

screen printing：丝网印刷

seal：黏合处，接缝处

seal the deal：达成交易

seam rubbing：磨平针眼孔

seam：接合（处，线，缝）

seat：后跟部位

SEBS：是 styrene-ethylene/butylene-styrene 的缩写，指聚苯乙烯 – 聚（乙烯 / 丁烯）– 聚苯乙烯嵌段共聚物

self-stratifying polymer：自分层聚合物

semi-elastic：半弹性

semi-rigid thermoplastic material：半刚性热塑性材料

sensible：理智的，合理的，朴素而实用的

service style boots：军靴，制式靴

shading：阴影

shank section：腰窝部位

shank：胫，小腿

sharpening stone：磨刀石

shod：穿鞋（或靴），穿着……鞋的

shoe insert：鞋垫

shoe tree：鞋撑

shoes zoning：鞋子分区

shore A：邵尔 A（硬度计量）

side (half skin)：半月革

side to side：（沿背脊线）对称地

silk screen-printing：丝网印刷

sinew：松紧带

size stick：尺寸测量器，量脚尺

size-fitting series No.：尺码型号

sizing：涂胶，上浆

sketchbook：速写本，素描册

skipped stitch：跳线

skiving：片边

sling backs：后空式高跟鞋

slingbacks：后跨带高跟鞋

slip：滑倒

slip-ons（slip-on shoe）：便鞋，一脚蹬鞋，无扣便鞋

slipper：拖鞋

slip-resistant feature：防滑性能

slip-resistant：防滑的

smooth：光滑的，烫平

snap：四合扣

soft tissues：软组织

sole：前掌，脚底，鞋外底

solid last：整体楦

solid：实心的

solids content：固（体）含量

somatosensory feedback：体感反馈

sophisticated：复杂的，精致的

sort：（鞋的）分拣

spandex fiber：氨纶纤维

speculative：推测的

splashback：（水流）溅射

split：剖层革

splitting：片料，片剖，通片

sponge rubber or PE (polyethylene) foam：发泡橡胶或 PE（聚乙烯）发泡材料

spread：遮盖（缺陷）

spring：（前后）跷，（鞋样板）升跷，起跷，样跷

stacked leather：堆叠皮革

stain：着色

stakeholders：利益相关者

steady-state walking：稳定状态行走

steel toe：钢包头

step frequency：步频

step length：步长

stiffener：主跟，定型材料

stiletto heel：锥形细高跟，匕首跟

stilettos：细高跟鞋

stirrup：马镫

stitchdown veldtschoen welt construction：双线透缝沿条结构

stitch marking：点缝帮标志点

stitch：缝纫

stitchdown construction：帮脚外翻线缝结构

stitched seam：合缝线

straight or mild semi-curve-lasted shoe：直形或半弧形鞋

strap：带条，绊带

straw：稻草

stretch / strength directions：拉伸 / 强度方向

stretchability：可延伸性

stretchable：可拉伸的

strobel shoes：由闯楦工艺制成的鞋

studded rubber sole：有圆形凸起的橡胶底，镶钉橡胶底

styrene butadiene rubber（SBR）：丁苯橡胶

styrene：苯乙烯

suede：反绒革

summer fashion line：夏季时装系列

supply chain risk：供应链风险

sustainable：可持续的

swing arm cutting press：摇臂裁断机

symmetric：对称

synthetic：合成的

synthetics：合成材料

tabi：（大拇指单独分开的）日式厚底短袜

table knife cutter：台式裁断机

tabletop：桌面，台面

tack：绷帮钉

tacky：发黏的，黏性的

tail：尾部

talus：距骨

tan：鞣制

tarsus：跗骨；踝骨

technical drawing：技术图纸

telescopic last：伸缩式鞋楦

tendon：肌腱

terrain boots：远足靴

textiles：纺织品

texture：质感

the noblest component：最主要的部件

therapeutic intervention：治疗性干预措施

thermal insulation performance：保温性能

thermo adhesive：热固性胶黏剂

thermoplastic：热塑性的，热塑性塑料

thermoregulatory：体温调节的

thin：削薄

thong：（固定用）皮条，皮带

thread size：纱线尺寸

three-dimensional：三维

three-piece last：三节楦

through：中底

tibia：胫骨

tight to toe：紧靠边地

tissue paper：防潮纸

toe box：包头

toe bumper：（硫化鞋）前围条

toe cap：包头

toe deformity：脚趾畸形

toe lift：（鞋的）前跷

toe peg sandals：夹趾凉（拖）鞋

toe tape：胶包头

tongue foam：鞋舌泡棉

tongue：鞋舌

tool temperature：模（具）温（度）

top line：鞋口，（靴子）筒口，楦头弧线

top piece：天皮，鞋跟面皮

top side：（部件的）上口边

topcoat：顶涂层

touch and close fastener：粘扣

track spike：田径鞋，跑鞋

traction：牵引，牵引力

traction：抓地力；牵引力

transducer：传感器，感应器

transverse arches：足横弓

traveling beam cutting press：动臂式裁断机

traveling head cutting press：动头式裁断机

tread patterns：鞋底花纹

tread：鞋底，外底面

treads and medallion stars：条带状和星状花纹

tricot：经编针织物

trim：修边

trim：装饰件，饰扣，（制革）修边，铣削，装饰用革

trip：绊倒

trunk：躯干

tuberosity：骨面粗隆，结节

turret-head machine：转塔头裁断机

twill：斜纹

ulcer：溃疡

ultraviolet yellowing：紫外线黄变

uncured rubber：未硫化橡胶

uncured：未硫化的，未固化的

underflap：帮脚，绷帮余量

underlay：可被遮盖的部位

underlie：构成……的基础

underslung：重心低的

undesirable finish：较差的光洁度

unit soles with raised walls：带边墙的成型底

upper closing：合帮套，帮面缝合

upper crimping：（靴）帮面压翘、压凹

upper cutting：帮面裁断

upper making / closing：制帮，帮部件的加工

upper：帮面

vamp lining：前帮衬里

vamp：前帮

vaquero：牛仔，牧人

vegetable-tanned leather：植鞣革

vibrating knife digital cutting machine：震动刀数字裁断机

viscosity：黏性，黏度

volatile organic compounds（VOC）：挥发性有机化合物

vulcanization：硫化

vulcanized rubber：硫化橡胶

vulcanized：硫化的

vulcanizing oven：硫化罐

waist：腰窝部位

walking pattern：行走模式，步态模式

walking velocity：步速

warp：经纱

water-based：水基型的，水性的

waterjet cutting：高压水束裁断

water-proof：防水的

water-resistant：拒水的

wax：打蜡

weak cementing：黏合不良

webbing：织带

wedge sole：楔形片底（前薄后厚）

wedge：楔状楦盖

weft：纬纱

weight-bearing：承重

welted shoes：沿条鞋

weltless：无沿条的

wiper plate：卡板

woven material：纺织材料

woven：编，织

wrap around construction：（帮料）包底结构

wrestling shoes：摔跤鞋

wrinkle：起皱，皱折